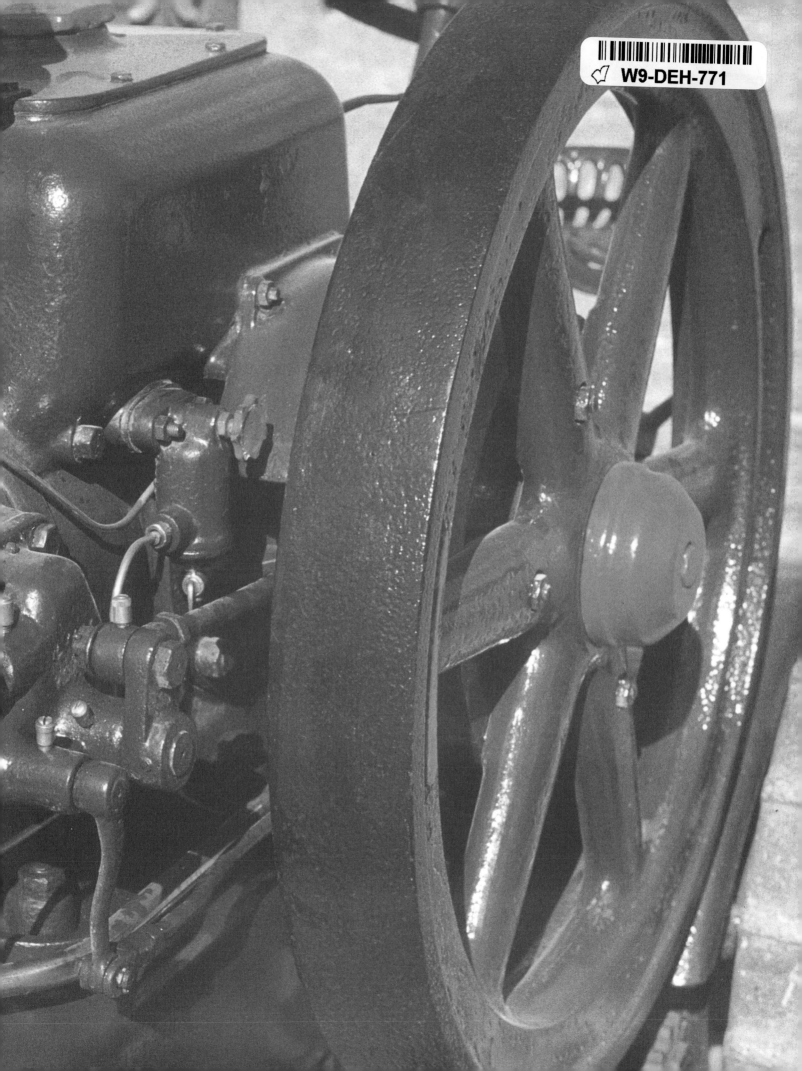

TRACTORS

THE WORLD'S GREATEST TRACTORS

TRACTORS

THE WORLD'S GREATEST TRACTORS

MICHAEL WILLIAMS

p

To Jayne

This is a Parragon Publishing book
First published in 2005

Parragon Publishing
Queen Street House
4 Queen Street
Bath BA1 4HE, UK

Copyright © Parragon 2005

ISBN 1-40545-320-6

Editorial and design by
Amber Books Ltd
Bradley's Close
74–77 White Lion Street
London N1 9PF
United Kingdom
www.amberbooks.co.uk

Project Editor: Michael Spilling
Design: Zoe Mellors
Picture Research: Natasha Jones

Printed in China

PICTURE CREDITS

All pictures courtesy of **Michael Williams** except the following:
Amber Books: 6, 7 (both), 14t, 15, 20t, 21, 27, 31l, 32 (main), 33, 34 (both), 37t, 41 (both),
49 (main), 54 (both), 65 (main);
David Williams: 10 (both), 12 (both), 13, 16, 17, 24t, 25, 38t, 43l, 45, 48, 49t, 52 (both), 56, 57,
59t, 60, 65t, 68, 70 (both), 79;
Peter Adams: 28t;
Andrew Morland: 31 (main), 35, 37 (main).

CONTENTS

INTRODUCTION

The development of tractors has had a massive impact on farming efficiency, and in developed countries tractor power replaced draft animals and the steam engine to revolutionize the way most of our food is produced. The first tractors arrived more than 100 years ago, and since then designs have improved to provide more power and efficiency. The important developments in tractor design are covered in this book, which describes more than 50 different models built in 10 countries.

The Early Years

Development started in the United States, and the list of models in this book starts with the 1892 Froelich, among the first tractors built in America. It was the earliest ancestor of the modern John Deere range, and was probably the first tractor with a reverse gear.

The demand for increased food production in Europe and North America during and just after World War I gave a big boost to tractor development. *Tractors* includes some important wartime models such as the Moline Universal, the most popular of the motor plows that enjoyed brief popularity at that time.

International Harvester's ruggedly reliable Titan 10-20 and the 8-16 Junior model that established the power take-off drive were also among the classics from this period.

Henry Ford's Fordson Model F was another wartime introduction. It remains the most successful tractor ever built, providing many thousands of farmers with affordable tractor power for the first time. A 1917 attempt to revive interest in steam power with the Garrett Suffolk Punch was a failure, but semi-diesel or hot bulb engines, pioneered by Lanz on their 1921 HL Bulldog, were the start of a highly successful tractor series that helped establish the semi-diesel on European farms.

Below: *Canadian-based Goold, Shapley, and Muir was one of a large number of tractor companies that failed to survive the intensely competitive era of the early 1920s.*

Left: *This cutaway picture shows the transmission that delivers the engine power on the big Steiger STX440 tractor from Case I.H.*

One of the most important developments of the 1930s was Harry Ferguson's three-point linkage implement mounting and control system. It became an international success on the Ford 9N tractor, followed by the 8N and the British-built Ferguson TE series.

Later Developments

Diesel power took a major step forward in the late 1940s and early 1950s, helped by the success of new easy-starting, smooth-running engines on a number of British-built tractors, including the New Fordson Major. The 1950s and 1960s were also periods of experiment and innovation that produced the amphibious County Sea Horse, the Doe Triple-D dual-engined model, and International Harvester's experimental gas turbine tractor.

Giant tractors were important from the 1970s onwards, and models featured in *Tractors* include the twin-engined Lely Multipower, the eight-wheel drive Versatile "Big Roy," Massey Ferguson's 4880, and the current Versatile 2425 from the Buhler factory in Canada. Recent developments featured are rubber tracks on the first Caterpillar Challenger, tractors with the latest constantly-variable transmission (CVT), and the Claas Xerion with its adjustable cab position.

The book also traces the increasing importance of driver comfort and safety—long

neglected by the tractor industry and its customers. Progress with safety cabs, noise reduction, suspension systems that offer a smoother ride, and electronic systems that simplify some operating routines have helped to provide a more acceptable working environment for those who spend long hours driving a tractor.

Below: *Rumeley OilPull tractors established a reputation as rugged heavyweights. An attempt to develop a new lightweight model range in the 1920s was less successful.*

FROELICH

1892 Froelich, Iowa, USA

FROELICH

Tractor history dates back to about 1889, when an American named John Charter used a gasoline engine mounted on the chassis and transmission of a steam-traction engine to power a threshing machine.

More experimental tractors appeared in 1892 and, like the Charter, they were used for threshing work. One of them was built by John Froelich of Froelich, Iowa, a village named after his father. John Froelich was a contractor who operated steam-powered threshing equipment in the Dakotas, and he decided to build a gasoline-powered traction engine to replace one of his steam engines.

Early Design
The chassis was built by a local blacksmith and was mounted on a set of traction-engine wheels. Froelich bought a Van Duzen engine, which was typical of the efficiency standards available from gasoline engines in the early 1890s. With just one enormous cylinder providing 2,155 cubic inches (35.5 liters) capacity it produced only 20hp (14.8kw).

Above: *This replica version of the original Froelich tractor was built by Deere and Co. to feature in a movie about the company's history.*

It was probably the first tractor with a reverse gear, which must have simplified the driver's job when maneuvering into position for threshing work, but there were few other concessions to driver comfort or safety. The driver stood right at the front of the tractor, and the only seat was the top of the fuel tank. The driving position was within easy reach of the engine's huge spoked flywheels, and the transmission gears were also exposed and easily accessible. There was, however, a large container for drinking water in case the driver was thirsty.

Waterloo Gasoline Traction Engine Co.

The tractor- or gasoline-powered traction engine worked well, and its success encouraged John Froelich and a group of businessmen to start a company to build more machines based on the same design. The company, based in Waterloo, Iowa, was called the Waterloo Gasoline Traction Engine Co., but the first two tractors built in 1893 failed to satisfy their customers and were returned to the factory.

Following this setback the company was reorganized in 1895. It stopped building tractors and concentrated on engine production instead, the name was changed to the Waterloo Gasoline Engine Co., and John Froelich left in 1895. But this was not the end of the story. Later, when it returned to tractor production, Waterloo was so successful that Deere & Co. bought the company in order to enter the tractor market—making the 1892 Froelich the earliest ancestor of all John Deere tractors.

Specifications

Manufacturer: John Froelich
Location: Froelich, Iowa
Model: N/A
Type: Self-propelled threshing engine
Power unit: Single-cylinder engine
Power output: 20hp (14.8kw)
Transmission: Exposed gears
Weight: N/A
Production started: 1892

Left: *Operator safety was a long way down the list of priorities when the Froelich tractor was built with its completely exposed flywheels and gearing.*

MARSHALL
🔧 **1908 Gainsborough, Lincolnshire, England**

MARSHALL 60HP

After a long and distinguished history as steam-engine manufacturers, the Marshall company of Gainsborough, Lincolnshire, in England, made its first move into the tractor market in 1907, when it announced its 30-hp model, powered by a paraffin-fueled engine.

Above: *This Marshall tractor was originally exported to Australia, but it recently made the return journey to Britain, where it has now been fully restored.*

Specifications

Manufacturer: Marshall Sons & Co.
Location: Gainsborough, Lincolnshire, England
Model: 60hp
Type: General purpose
Power unit: Marshall two-cylinder engine
Power output: 60hp (44.4kw)
Transmission: Three-speed gearbox
Weight: 22,000lb (9,988kg)
Production started: 1908

The new model, weighing about 4.9 tons (5 tonnes), was big by British standards and was designed mainly for export. Marshall targeted countries with big farms such as Canada and Australia, where its steam engines were already well established and had built up a name for long-term reliability.

An additional model followed in 1908, powered by a rear-mounted, four-cylinder engine consisting of a pair of its two-cylinder 30-hp (22.2-kw) engines mounted side by side to give a 60hp (44.4kw) output. The cylinder bore and stroke were both 7in (17.8cm) and the rated engine speed was 800rpm.

Mixed Fortunes

The Canadian market was dominated by American-built steam engines and tractors. Marshall competed strongly, entering its 30- and 60-hp tractors plus a steam-traction engine in the Agricultural Motor Competition held in Winnipeg, Manitoba, in 1909.

The results were disappointing, as the 30-hp tractor came third in a class of three and the steam engine was withdrawn. The new 60-hp tractor came second in a class of five, however, and in the same year it also achieved valuable publicity by winning the Gold Medal award at the prestigious Brandon Fair, in Manitoba.

Above, left: *Marshall was one of several British steam-engine manufacturers that built big, heavyweight tractors for export to far corners of the British Empire.*

SAUNDERSON
✖ **1910 approx   Elstow, Bedford, England**

SAUNDERSON UNIVERSAL MODEL G

Below: There was enough space in front of the radiator on the Saunderson Universal tractor for a wooden box to carry tools, spare spark plugs, and other essentials.

Britain's urgent need for tractor power to increase food production during World War I affected even the Royal Family. There was plenty of scope to plow more land on the royal estate at Sandringham in Norfolk, and when King George V decided to buy a tractor he chose a Saunderson Model G.

Specifications

Manufacturer: Saunderson

Location: Elstow, Bedford, England

Model: Universal G

Type: General purpose

Power unit: Two-cylinder engine

Power output: 20hp (14.8kw)

Transmission: Three-speed gearbox

Weight: N/A

Production started: 1910 approx

The order was placed in 1916 and it took two days to drive the tractor the 80 miles (129km) from the factory to the Royal estate. The company gained valuable publicity from the order and from the subsequent reports about the tractor's progress on the Royal estate, and there was more publicity from the Saunderson family farm, probably the first in the country to be farmed entirely with tractor power. Other successes with the Model G included an agreement for a French company to build the tractor under license.

Saundersons

H. P. Saunderson, a baker's son, started building tractors in 1900. The Model G was available from about 1910, powered by a Saunderson-designed two-cylinder vertical engine that was described as 20hp (14.8kw) and later as 25hp (18.5kw).

After the war Saunderson faced financial problems and in 1924 the Manchester-based Crossley engineering company bought the Saunderson company and attempted to market the tractors for road haulage, but production ended within about two years.

MOLINE
⚒ 1914 Moline, Illinois, USA

MOLINE UNIVERSAL MOTOR PLOW

The motor plow was very definitely designed to replace horses or mules for pulling cultivators and other machinery, and between about 1915 and 1920 they sold in large numbers in North America and, to a lesser extent, in several European countries.

Although they were slow, often under-powered and acquired a reputation for being awkward to use and unstable, motor plows were a relatively low-cost alternative to a conventional tractor, and they offered many thousands of small-acreage farmers their first realistic opportunity to make the switch from animal power to power farming. Motor plows were built by some of the big tractor companies in the USA, but they were also manufactured by many of the small businesses that mushroomed into the tractor market during the World War I sales boom.

The most successful and probably the best designed of the American motor plows was the Moline Universal. The first version was built in 1914 by the Universal Tractor Manufacturing Co. of Columbus, Ohio, but in the following

Above: Motor plows enjoyed a surge of popularity during World War I, and the Moline Universal was the leading model.

year the design was bought by the Moline Plow Co. of Moline, Illinois, and Moline introduced a number of improvements.

The power unit for the original Universal was a two-cylinder engine built by the Reliable company, and this version remained in production after the Moline takeover. Moline also introduced an additional version called the Model D, which was powered by a four-cylinder engine. Model D production started with a bought-in engine, but in 1917 this was replaced by an engine designed and built by the Moline company.

Advanced Features

The cylinder dimensions of the Moline engine were 3.5-in (8.9-cm) bore and a 5.0-in (12.7-cm) stroke, and the designers also added advanced features from the car industry, including an electric governor. In 1918 the Universal was the first tractor equipped with an electric starter, and a headlight was also included in the equipment list. Other design improvements introduced on the Model D included a heavy concrete disk attached to each driving wheel to provide extra weight for improved stability.

Demand for motor plows faded rapidly after about 1919, and production of the Moline Universal version ended in the early 1920s. In 1929 the Moline Plow Co., then known as the Moline Implement Co., was one of three companies involved in the merger to form the Minneapolis-Moline Power Implement Co.

Specifications

Manufacturer: Moline Plow Co.	**Power output:** 18hp (13.32kw)
Location: Moline, Illinois	**Transmission:** N/A
Model: D	**Weight:** 3,590lb (1,630kg)
Type: Motor plow	**Production started:** 1914
Power unit: Moline four-cylinder engine	

Left: *The Moline Universal was available with a choice of engines, and in 1918 it became the first tractor to be equipped with an electric starter motor.*

INTERNATIONAL HARVESTER
1915 Milwaukee, Wisconsin, USA

INTERNATIONAL HARVESTER TITAN 10-20

For much of the first 50 years of tractor history International Harvester was the most successful manufacturer worldwide, and the Titan 10-20 model was among the most important and successful tractors the company built.

Production started in 1915 and, when the last of the 10-20s rolled off the production line at the I.H. factory in Milwaukee, Wisconsin, the build total was approaching 80,000.

Reliability
Although they looked quite different, the Titan and Mogul models shared a number of

important features. Both were ruggedly built and, at a time when some manufacturers were experimenting with new ideas, the engineers at International Harvester chose a more traditional approach based on a steel-girder frame and a heavy, slow-revving two-cylinder engine. The result, by 1915 standards, was a reputation for reliability that helped International Harvester

Above: The Titan 10-20 was another of the lightweights that helped to make International Harvester the leading tractor company before the arrival of the Fordson.

to take the lead in the small to medium sector of the market.

The cylinders of the 10-20 engine were horizontal with the crankshaft at the front and the cylinder head just in front of the driver. Cylinder bore was 6.5in (16.5cm) with 8-in (20.3-cm) stroke, and the 20-hp (14.8-kw) rated output was produced at 500rpm. The fuel was kerosene, but an injection device added water to control the temperature and prevent pre-ignition. The ignition system was based on a high-tension magneto with an impulse starter.

Cooling System

The big 39-gallon (117-liter) capacity cylindrical tank located over the front axle was the cooling system, which matched the general simplicity of the 10-20, relying on heat difference and a steam impulse to circulate the water to and from the engine and avoiding the need for pumps and cooling fans. It was a system that was already becoming dated when it appeared on the Titan, but it worked well and contributed to the 10-20's reliability.

A two-speed gearbox provided a top gear speed of 2.8mph (4.5km/h) forward and in reverse, with the final drive through two exposed chain-and-sprocket drives. The weight of the 10-20 was 5,225lb (2,372kg), making it the lightest model in the Titan series, and the overall height was only 67in (170cm) compared with from 110in (279cm) and higher for other Titan models.

Few design changes were made during the production life of the 10-20, but the obvious change was a switch to bigger rear fenders, or mudguards, introduced in 1919.

Specifications

Manufacturer: International Harvester Co.
Location: Milwaukee, Wisconsin,
Model: Titan 10-20
Type: General purpose
Power unit: Two-cylinder horizontal engine
Power output: 20hp (14.8kw)

Transmission: Two-speed gearbox
Weight: 5,225lb (2,372kg)
Production started: 1915

Below: *Engine cooling for the Titan 10-20 was based on a water supply in the big cylindrical tank over the front wheels; circulation was based on temperature differences.*

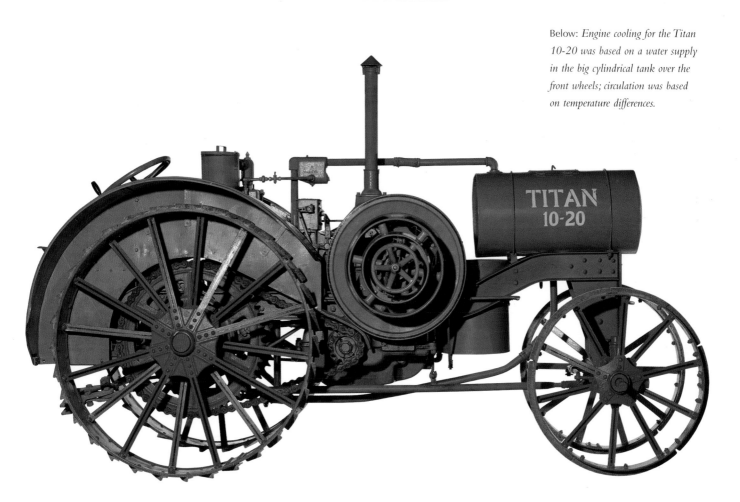

CLAYTON
⚒ **1916 Lincoln, Lincolnshire, England**

CLAYTON

For a brief period during and immediately after World War I British farmers showed increased interest in crawler tractors, and it has been suggested that this was to some extent encouraged by the enormous publicity that followed the success of Britain's tracklaying tanks as they coped with difficult battlefield conditions.

Several UK-based companies moved into the crawler-tractor market to compete with Caterpillar, Cletrac, and other imports from North America. The most successful was the Clayton, which was made by the Lincoln-based Clayton and Shuttleworth company. The Clayton tractor was available from about 1916, and the first two years of production were boosted by the UK government, which placed substantial orders for the tractors for use in wartime plowing-up campaign.

War Service
The Clayton in the photograph above, dating from 1918, has a War Department emblem stamped on the brass serial-number plate,

Above: This is one of a batch of Clayton tracklayers that were ordered by the British government for use in the wartime plowing campaign.

indicating that it was probably one of the original government tractors.

The Clayton was basically a simple design, with the fuel tank distinctively mounted above the engine compartment. It featured a steering wheel instead of the more usual steering levers, and the steering mechanism operated through two fully exposed cone clutches—one on each side—that controlled the drive to the tracks. When sharp turns were needed the driver could use foot-operated brakes to lock one of the tracks, forcing the tractor to swivel around in little more than its own length.

Engine

Most of the Clayton tractors were equipped with a four-cylinder Dorman engine adapted to run on paraffin, but Aster engines were also available. The output from the Dorman power unit was described as 35hp (26kw), but toward the end of the production run this increased to 40hp (29.8kw). A two-speed gearbox provided a top speed of 4mph (6.6km/h), with 3mph (4.8km/h) available in reverse.

Clayton continued to build the crawler tractor until the mid-1920s when production was temporarily suspended. It was reintroduced again in 1928, but this was also a short-lived arrangement because Clayton and Shuttleworth was taken over in 1930 by Marshall of Gainsborough.

All the tractor development activity carried out at Marshall in the early 1930s was concentrated on its single-cylinder diesel, and production of the Clayton crawler tractor ended in about 1931.

Development

It seems, with the benefit of hindsight, that Marshall should have made the decision to put some resources into improving the Clayton tractor. When the Marshall company took over, the Clayton was a well-established design with a good reputation, even if it did need updating. A combination of the single-cylinder Marshall engines plus the Clayton would have provided an impressive tractor range for Marshall to sell in the 1930s.

Specifications

Manufacturer: Clayton and Shuttleworth

Location: Lincoln, Lincolnshire, UK

Model: Clayton

Type: Tracklayer

Power unit: Dorman four-cylinder engine

Power output: 35hp (26kw)

Transmission: Two-speed gearbox

Weight: not known

Production started: c. 1916

Left: *Driver comfort and safety were a low priority when the Clayton tractor was designed, with the steel trackplates and the belt pulley just a few inches from the driver's feet.*

WATERLOO BOY
✗ 1917   Waterloo, Iowa, USA

WATERLOO BOY MODEL N

The Waterloo Gasoline Engine Co. was established in 1895 as an engine manufacturer following problems encountered when selling a production version of John Froelich's 1892 tractor. The engine business was a success, and in 1912 the company moved back into tractor production to provide an additional outlet for its engines.

For their 1912 tractor the designers had chosen a four-cylinder engine, but this was replaced by a horizontal two-cylinder design when the Waterloo Boy L and R models arrived in 1914. The two-cylinder engine was probably more reliable and cost less to produce than a four-cylinder; it was also the start of America's best-known tractor engine series.

The engine for the Model R had 5.5-in (13.9-cm) bore and 7-in (17.7-cm) stroke, but 0.5in (1.3cm) was added to the bore in 1915, with a further 0.5in (1.3cm) in 1917 when the new Model N arrived. The Model N was also equipped with a two-speed gearbox instead of

Above: The twin-cylinder horizontal engine on later versions of the Waterloo Boy tractor was adopted for John Deere tractors.

the single speed on earlier versions, and another improvement was fitting roller bearings instead of the plain bearings used previously.

Ferguson and Deere & Co.

Both the Model R and the later Model N were successful, and they had an unexpected influence on future developments in the tractor industry. Waterloo Boy tractors were exported to Britain during the war years, where they were known as the Overtime, and the distributor for what is now Northern Ireland was a Belfast garage-owner named Harry Ferguson. Ferguson personally supervised some Overtime demonstrations, and this may have influenced the ideas that produced his Ferguson System of implement attachment and control.

Waterloo Boy tractors also attracted the interest of Deere & Co. At that time Deere was a highly successful machinery manufacturer with ambitions to move into the fast-expanding tractor market, but its attempts to develop a new John Deere tractor had met with little success. Purchasing an existing tractor company offered a shortcut into the market with an established product, and buying the Waterloo Boy company would give Deere & Co. its own engine facilities as well.

The takeover was completed in 1918 when Deere paid $2.35 million for the whole Waterloo company. The Model N tractor remained in production with minor modifications and still carrying the Waterloo Boy name until 1923.

Specifications

Manufacturer: Waterloo Gasoline Engine Co.

Location: Waterloo, Iowa

Model: N

Type: General purpose

Power unit: Two-cylinder horizontal engine

Power output: (rated) 25hp (18.5kw)

Transmission: Two-speed gearbox

Weight: 6,183lb (2,807kg)

Production started: 1917

Left: *The final drive on the Waterloo Boy tractors was exposed to dust and mud, but this did not prevent the tractors from establishing a reputation for reliability.*

Left: *The Waterloo Boy badge on the fuel tank was removed from tractors exported to Britain, where they were sold under the Overtime name.*

FORDSON
✖ 1917 Dearborn, Michigan, USA

FORDSON MODEL F

Although Henry Ford's first experimental tractor was completed in 1906 or 1907, the project remained a low priority for a few years because the success of his cars filled most of his time.

More experimental models emerged from time to time, proving that Ford had not abandoned his idea of developing a mass-produced tractor that small farms could afford, but it was World War I that put more urgency behind the development work.

Separate Tractor Company

In 1915 Henry Ford formed a separate company in Dearborn, Michigan, to develop the tractor project. It was called Henry Ford & Son and it employed some of Ford's most talented engineers, and by 1916 Ford gave the go-ahead for a batch of about 50 pre-production prototype tractors to be built for evaluation. Two were shipped to Britain at the request of the government to test their suitability for the wartime plowing campaign.

The tractors arrived in January 1917, and experts who observed the tests liked their

Above: The Model F was the biggest-selling tractor ever built, helped by Henry Ford's price-cutting policy, which put many rival manufacturers out of business.

performance and light weight. The fact that these were the first tractors designed specifically for mass production to reduce production costs also attracted the British government, and it asked Henry Ford to supply 6,000 tractors to Britain as quickly as possible.

Although Ford wanted to do more development work, he agreed to the British request and started preparations for production. The tractor was the Fordson Model F, and the production total reached 254 by the end of the year, peaking at more than 100,000 per year in the mid-1920s. More than 700,000 Fordsons were built in Dearborn before production was switched to Ireland in 1928.

Design Features

The most advanced design feature of the Model F was the way the housings for the engine, transmission and differential were brought together to form a fully enclosed, rigid unit that eliminated the need for a separate frame. Power

was provided by a four-cylinder engine with a water-washer air cleaner, a worm gear provided the final drive, and there were no brakes on the early versions.

The huge production volumes were helped by bulk orders from the new Communist government in Russia. With an urgent need to increase the amount of mechanization on Russian farms in order to boost food production, the Russians imported more than 26,000 Fordsons and also built an unknown number under license in Russia. Henry Ford's price-cutting policy also boosted Model F sales. The Fordson list price in America was $750 in 1918, but this was cut repeatedly to reach $395 in 1922.

Specifications

Manufacturer: Henry Ford & Son
Location: Dearborn, Michigan
Model: F
Type: General purpose
Power unit: Four-cylinder engine
Power output: 18hp (13.3kw)

Transmission: Three-speed gearbox
Weight: 2,710lb (1,230kg)
Production started: 1917

Below: *This side view shows how the Model F engine, gearbox, and back axle were joined together in one rigid dirt-proof structure.*

GARRETT
✚ 1917   Leiston, Suffolk, England

GARRETT SUFFOLK PUNCH

The Garrett company began making horse-drawn implements in Leiston, Suffolk, in 1782, and by the time it celebrated its first centenary it had already adopted the latest technology and was building agricultural steam engines.

Garrett became one of the biggest manufacturers of steam engines in Britain, with a successful export business, and when tractors provided the next power-farming revolution, it responded to the challenge in an unusual way.

Many of the old-established steam-engine companies tried to ignore the threat posed by the new competitors, but some adopted a more progressive approach and switched to making tractors. Garrett decided to redesign the traditional steam-traction engine and turn it into a new steam-powered tractor. It called it the Suffolk Punch after a well-known heavy horse breed—not, perhaps, the most suitable image for a tractor designed for the future—and it was developed to eliminate some of the disadvantages of the traditional traction engine.

Above: The Suffolk Punch steam tractor was Garrett's ambitious attempt to modernize the steam-traction engine, but it could not compete with low-cost tractor power.

Design Features

The driver was moved from his usual platform at the rear to a seat at the front, with a big improvement in forward visibility. The boiler was moved to the rear, putting the weight over the driving wheels to aid traction, and this also reduced the weight over the steering wheels at the front, allowing a more precise, driver-friendly Ackermann steering mechanism to replace the traditional chain-operated steering gear. There was also a suspension system over the tractor's rear axle.

As well as moving the boiler to the rear, the design team also located an auxiliary water tank over the rear axle to put additional weight on the driving wheels. This was because the tractor was designed to pull plows and other implements in the field instead of providing stationary power for threshing or for a cable-operated plowing system. It was a major step forward because steam-powered direct traction for cultivations was extremely unusual in Britain, where the soil is often heavy and easily damaged by a heavy traction engine.

Engine

The Suffolk Punch power unit was a double-crank compound engine fueled by coal and producing 40hp (29.6kw), and the first plowing tests in 1917 were encouraging. There was plenty of optimism about the steam tractor's future, but this turned to concern, according to one report, when performance and cost figures arrived for the new Fordson tractor.

Although the Suffolk Punch was comfortably ahead on plowing performance, the economic advantage heavily favored the Fordson. Garrett built eight of the Suffolk Punch tractors, and the sole survivor is in a museum in the factory where it was built.

Specifications

Manufacturer: Richard Garrett & Sons

Location: Leiston, Suffolk, England

Model: Suffolk Punch

Type: Steam tractor

Power unit: Double-crank compound steam engine

Power output: 40hp (29.6kw)

Transmission: N/A

Weight: N/A

Production started: 1917

Below: *Garrett engineers redesigned the traditional traction engine, placing its driver at the front and the boiler at the rear of the Suffolk Punch.*

INTERNATIONAL HARVESTER
1917 Chicago, Illinois, USA

INTERNATIONAL HARVESTER 8-16

International Harvester's commitment to the lightweight sector of the tractor market took a big step forward with the launch of the 8-16 or Junior model in 1917. It was built at the Chicago factory, the home of the I.H. Mogul series tractors, but I.H. had already taken the decision to phase out the Mogul and Titan names, and the 8-16 was simply sold as an International Harvester tractor.

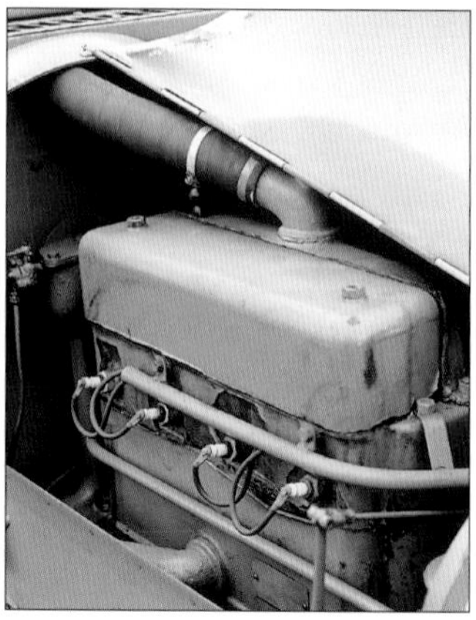

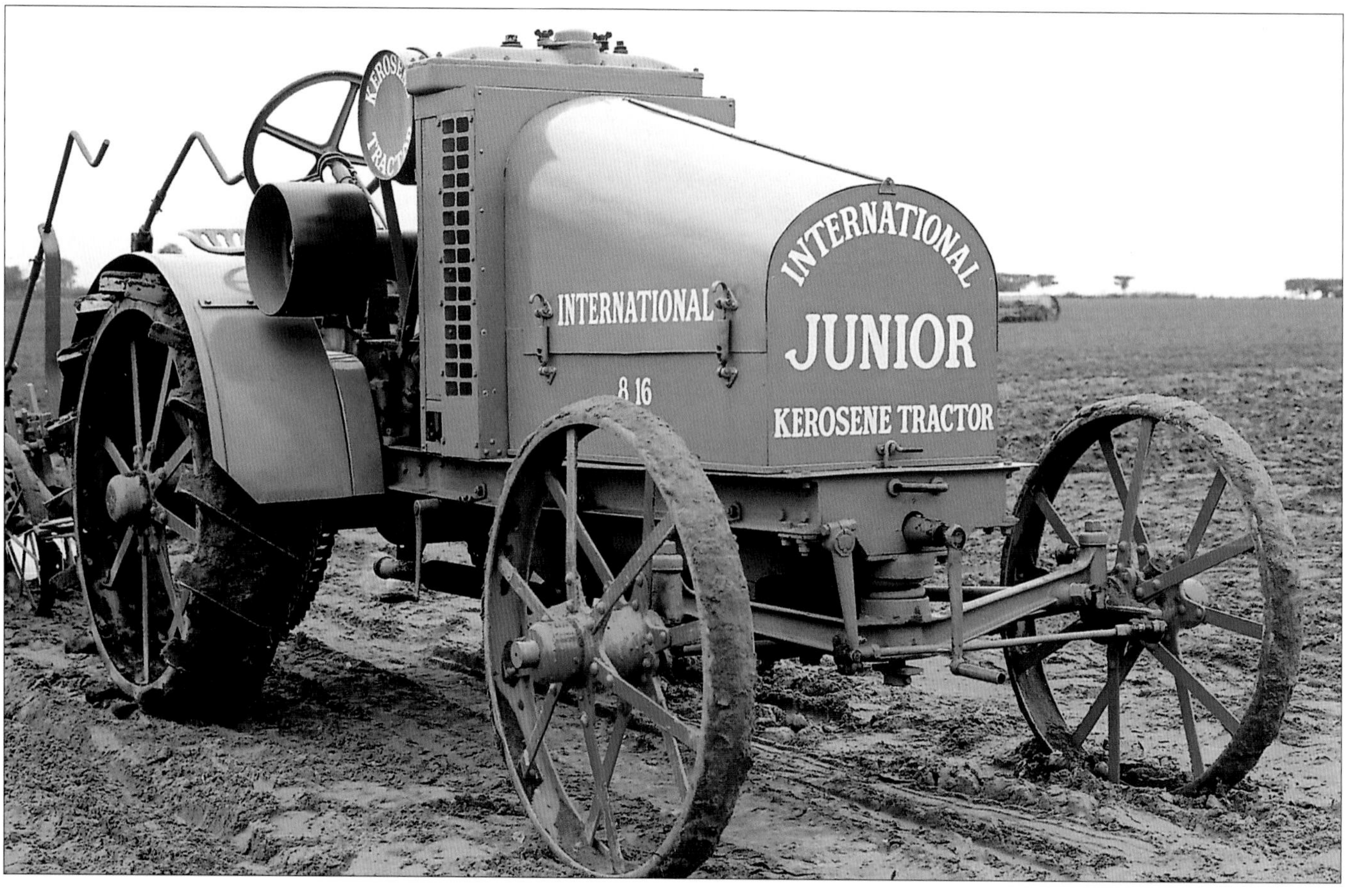

The 8-16 arrived at a time when tractor design was progressing rapidly, and the new model had a number of advanced features. One of these was the sharply downward slope of the hood, ensuring good forward visibility from the driver's seat for increased steering accuracy. To achieve this downward slope the International Harvester engineers had to move the radiator from its traditional location at the front of the tractor to a new position behind the engine, a layout that was also adopted on the French-built Renault tractors at that time. Another 70 years

Above, top: *International Harvester used what was then a more modern four-cylinder engine for the 8-16.*
Above: *To achieve the sloping hood line of the 8-16, the radiator was moved behind the engine.*

were to pass before the sloping hood line was generally adopted by the tractor industry.

I.H. chose a four-cylinder engine for the 8-16, a step up from the twin-cylinder power unit in the earlier Mogul 8-16 and the 10-20 Titan available at the same time. The bore and stroke measurements of the new engine were 4.0in and 5.0in (10.1cm and 12.7cm), and the rated speed was 1,000rpm.

Another feature that put the 8-16 ahead of its rivals was the power takeoff shaft. This was not an I.H. invention, as the Scott tractor from Scotland was equipped with a similar device in 1904, but it was International Harvester with its 8-16 tractor that established the power takeoff as an important permanent addition to tractor versatility and efficiency.

Sales Success

Its small size and light weight, plus the addition of a power takeoff, made the 8-16 one of the most versatile and advanced tractors available, and its sales success extended to the United Kingdom where some of the tractors had been imported to help with the wartime plowing campaign. They were well suited to UK conditions and established a good reputation for performance and reliability.

Although the 8-16 design included some advanced features, the exposed chain-and-sprocket final drive with no protection from dirt and mud was becoming outdated by 1917. It was certainly out of date by 1922 when International brought the 8-16's five-year production run to an end.

Specifications

Manufacturer: International Harvester
Location: Chicago, Illinois
Model: 8-16
Type: General purpose
Power unit: Four-cylinder engine
Power output: 16hp (12kw)
Transmission: Three-speed gearbox
Weight: 3,660lb (1,662kg)
Production started: 1917

Left: *A metal seat mounted on a piece of spring steel was typical of the driver comfort provided on early 1920s tractors.*

MASSEY-HARRIS

c. 1919 **Weston, Ontario, Canada**

MASSEY-HARRIS NO. 3

When Massey-Harris signed an agreement to build the Parrett tractor at its factory near Toronto, it was the company's second attempt to move into the lucrative and fast-expanding agricultural tractor market.

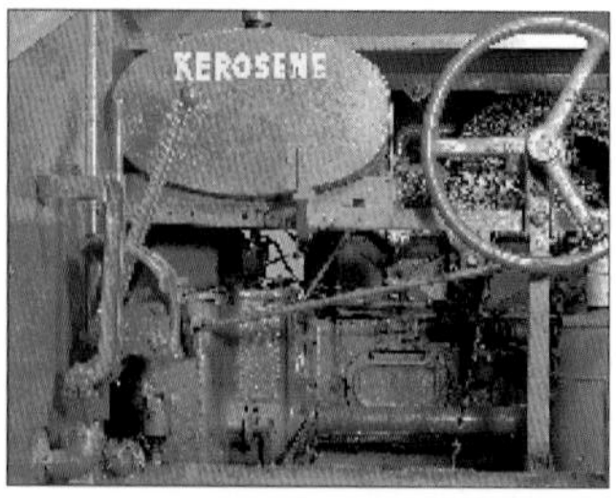

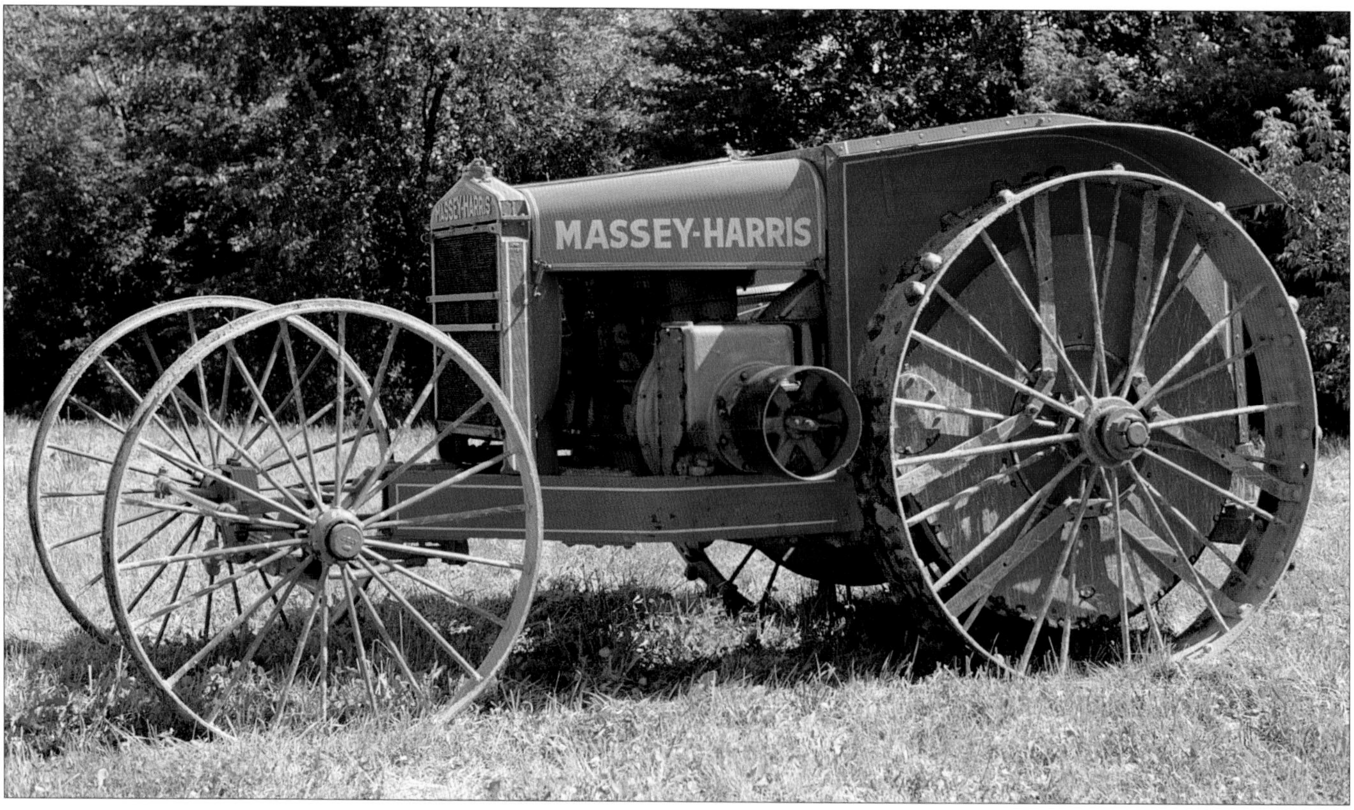

An earlier agreement to distribute the Big Bull tractor in Canada ended when the American manufacturer failed to make delivery of the tractors because of production problems. For their next attempt Massey-Harris decided to control production as well as sales, and the tractor it chose was designed by the Parrett brothers in Chicago.

The agreement was signed in 1918, with production starting the following year, and the tractors carried the Massey-Harris name and were sold in Canada through the extensive Massey-Harris dealer network.

Big Wheels

A distinctive feature of the Parrett was the 4-ft (1.2-m) diameter front wheels, and the Parrett brothers had plenty of reasons why big wheels were better than small ones, even if they did look frail. Large-circumference wheels rode more easily over obstructions, and they also spread the weight of the tractor over a bigger area to reduce soil compaction, the Parretts claimed.

Wheel bearings also lasted longer, they said. This is partly because a big wheel rotates more slowly than a small one when the tractor is moving, and the bearings are also further from

Above, top: *The spacious platform had all of the controls within easy reach of the operator. The steering wheel was connected by a chain-and-sprocket device to the front end.*

Above: *Large-diameter front wheels inherited from the original Parrett design were an important selling feature for the new Massey-Harris tractor.*

the ground and away from the damaging effects of dust and mud.

Sales of Massey-Harris–built Parrett tractors made an encouraging start, but it is not clear if Canadian farmers were impressed by the sales pitch for the big wheels or were simply reassured by the familiarity of the Massey-Harris brand name.

Versions

Massey-Harris offered three versions of the Parrett tractor. Numbers 1 and 2 were basically similar with a 12-25 power rating—(12hp) 8.9kw at the drawbar and 25hp (18.6kw) on the belt pulley—and the radiator on both was sideways facing. Both featured a two-speed gearbox allowing speeds of 2.4mph and 4.0mph (3.8km/h and 6.4km/h), but this was considered too fast and the maximum speeds

were reduced to 1.75mph and 2.4mph (2.8km/h and 3.8km/h) when the third model was introduced.

Other changes on Massey-Harris No. 3 included increasing the cylinder bore and stroke to 4.5in and 6.5in (11.4cm and 16.5cm), respectively, to boost the power rating to 15-28, and the radiator was turned through 90° to the more conventional forward-facing position.

Meanwhile shrinking demand for tractors together with intense competition plus Henry Ford's price-cutting policy were causing problems in the industry. The Parrett company stopped building tractors in about 1922 and Massey-Harris ended its tractor production in the following year.

Specifications

Manufacturer: Massey-Harris
Location: Weston, Ontario, Canada
Model: No. 3
Type: General purpose
Power unit: Four-cylinder engine
Power output: 28hp (20.72kw)

Transmission: Two-speed gearbox
Weight: N/A
Production started: c. 1919

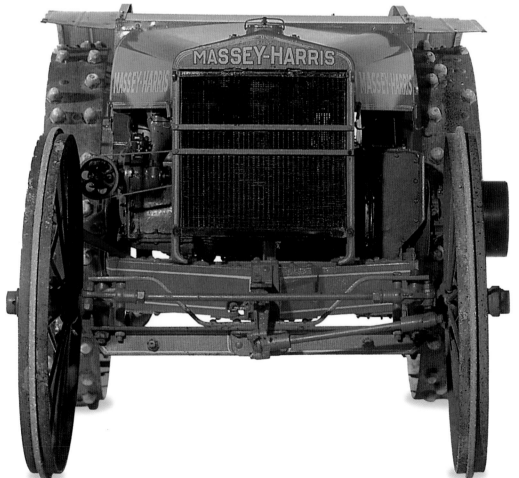

Left: *One of the distinguishing features of the No. 3 tractor was the forward-facing radiator, unlike the sideway-facing position on the Massey-Harris 1 and 2 models.*

FIAT
✻ 1919 Turin, Italy

FIAT 702

When World War I ended in 1918 tractors were urgently needed to increase Europe's food production, and in Europe, this brought a big batch of new companies into the industry.

Above: *Fiat was one of the big European car and truck companies that decided to move into tractor production after the end of the war.*

Left: *Fiat 702 drivers had the luxury of a transverse leaf-spring suspension over the front axle, a feature that was probably borrowed from Fiat cars and trucks.*

Fiat was one of the first big companies to begin tractor production in Europe. The Fiat car- and truck-making history stretches back to 1899, but tractor production did not start until 20 years later when the 702 model arrived. It was available as a standard agricultural version, and there was also an industrial version equipped with solid rubber tires.

Design Features

Fiat chose a four-cylinder gasoline/paraffin engine for the 702. It developed about 25hp (18.5kw) and the power was delivered through a three-speed gearbox that also provided three driving speeds for the belt pulley. The pulley was mounted at the rear, close to the driver's seat, which was offset to allow better forward visibility. An unusual feature was the transverse leaf spring over the front axle to give a smoother ride.

This was probably a result of Fiat's experience with designing other vehicles—suspension systems to improve driver comfort were standard equipment on cars and commercial vehicles almost 100 years before they were widely available on farm tractors.

The 702 was replaced in about 1921 by the 703 model with an improved transmission, and 702 owners were offered an update kit to bring their old tractors up to the latest specification.

Specifications

Manufacturer: Fiat
Location: Turin, Italy
Model: 702
Type: General purpose
Power unit: Four-cylinder engine
Power output: 25hp (18.5kw)
Transmission: Three-speed gearbox
Weight: 5,720lb (2,600kg)
Production started: 1919

RENAULT
1919 **Billancourt, France**

RENAULT GP

The list of European companies moving into the tractor market for the first time in the postwar period includes Citroën, Peugeot, and Renault, all of the big three companies in the French motor industry. It was Renault that had the biggest advantage initially because it was able to base the design of its first tractor on a small tank it had developed during World War I for the French army.

Below: *Renault based the design of the GP crawler tractor on a light tank it had built for the French army during the war.*

Specifications

Manufacturer: Renault	
Location: Billancourt, France	
Model: GP	
Type: Tracklayer	
Power unit: Four-cylinder engine	
Power output: 30hp (22.3kw)	
Transmission: Three-speed gearbox	
Weight: N/A	
Production started: 1919	

The new crawler tractor arrived in 1919 and was known as the GP. The power unit was a four-cylinder gasoline engine developing about 30hp (22.3kw); the transmission had a cone clutch linked to a three-speed gearbox.

A distinctive design feature inherited from the tank was the sloping hood line that helped to give improved forward visibility from the driver's seat. This was achieved by positioning the radiator between the engine and the driver, with the fuel tank mounted above the radiator. Another feature inherited from the wartime Renault tank was the tiller steering to change the direction of the tractor by controlling the drive to the tracks. When an improved version of the GP, known as the H1, was introduced in about 1920, the tiller steering was retained, but the addition of a pair of bicycle-style handlebars made the tiller easier to operate.

Indications for Future

The GP tracklayer was aimed at the top sector of the French market and was not a big-selling tractor, but it achieved sufficient success to encourage Renault to develop new crawler and wheeled models based on the same design.

LANZ
✖ 1921 Mannheim, Germany

LANZ HL BULLDOG

When the first of the Lanz Bulldog tractors arrived in 1921 it was the start of what was to become one of the longest and most influential tractor series Europe has produced.

Specifications

Manufacturer: Heinrich Lanz

Location: Mannheim, Germany

Model: HL Bulldog

Type: Designed for belt work

Power unit: Single-cylinder hot bulb engine

Power output: 12hp (9kw)

Transmission: Single-speed gearbox

Weight: N/A

Production started: 1921

The Heinrich Lanz company was based in Mannheim, where it started making farm machinery during the 1860s. Production started with a range of stationary equipment, such as straw choppers and small threshing machines, and these were followed later by steam engines. But it was the introduction of the Bulldog tractors that had the biggest impact on the company's future development.

Design Features

The engineer who designed the first of the Bulldog tractors was Dr. Fritz Huber, who had joined the company in 1916. The HL Bulldog was his first production tractor, based on a single-cylinder semidiesel engine mounted on a self-propelled chassis. The engine output was 12hp (9kw), and the lack of ground clearance and the relatively small-diameter wheels show that the Bulldog was designed to power stationary equipment rather than do field work itself. Solid rubber tires were offered as an option, and the standard specification included foot-operated brakes with wooden blocks acting on both rear wheel rims.

Sales of the HL Bulldog during the eight-year production run totaled just over 6,000, but by 1956 the production total for Bulldogs of all models had reached more than 200,000, and large numbers of Bulldog lookalikes had been built in other European countries under various licensing agreements.

Above: The classic designs of tractor history come in all shapes and sizes, including the little HL Bulldog tractor from Lanz, the first of a long line of Bulldog tractors.

CASE

✕ 1922 Racine, Wisconsin, USA

CASE 12-20

Case Crossmount series tractors with their transverse engine layout covered a wide range of power outputs, including the little 12-20 model, which was available starting in 1922. It was a classic Crossmount tractor, complete with a sturdy one-piece cast-iron frame; however, the pressed steel front and rear wheels were a distinctive 12-20 feature.

Right: *Case's 12-20 shared most of the Crossmount design features, but the pressed-steel front and rear wheels appeared only on the 12-20.*

Above: *This rear view of the little 12-20 shows the offset driving position—it also suggests that a repair job is overdue to deal with the damaged exhaust stack.*

Specifications

Manufacturer: J. I. Case Threshing Machine Co.

Location: Racine, Wisconsin

Model: 12-20

Type: General purpose

Power unit: Four-cylinder engine mounted tranversely

Power output: 22.5hp (16.7kw) (maximum)

Transmission: Two-speed gearbox

Weight: 4,450lb (2,020kg)

Production started: 1922

Case introduced the 12-20 as a replacement for the 10-20 three-wheeler model, and it remained in the range until about 1929. By that time Case had changed its system of model identifications, using letters of the alphabet instead of the rated power figures, and when the 12-20 ended its production life it was known as the Model A.

Engine Features

The cross-mounted engine was designed and built by Case and was a four-cylinder vertical design with overhead valves. Cylinder bore was 4.1in (10.4cm) with 5.0-in (12.7-cm) stroke. The rated speed was 1,050rpm, and a special feature of the engine design was the use of replaceable cylinder sleeves to simplify maintenance. The maximum engine power when the 12-20 was tested in Nebraska was 22.5hp (16.7kw), but this reduced to 20.17hp (15kw) when the engine was tested at its rated speed.

The 12-20 was one of the last of the distinctively styled Case Crossmount tractors, and when the replacement models arrived they were designed with the engine in the normal lengthwise position.

JOHN DEERE

⚒ 1923 Waterloo, Iowa, USA

JOHN DEERE MODEL D

After taking over the Waterloo Boy company, Deere & Co. continued to build Model N tractors with just a few minor improvements, and it also continued to use the Waterloo Boy name.

Waterloo Boy engineers were already working on the replacement for their Model N tractor in 1918 when their company was taken over, and the project continued under the management of the new owners.

When the replacement arrived in 1923 it was known as the Model D, and it was the first tractor to carry the John Deere name and to sell in significant numbers.

New Features
Almost every feature of the old Waterloo Boy design was replaced on the new Model D, but

the principle exception to this was the tractor's two-cylinder horizontal engine.

There was an increase in the maximum power at 30.4hp (22.5kw) compared with the Model N version's 26hp (19.4kw), and the engine speed was given a 50rpm boost to 800rpm for the new tractor, but the bore and stroke measurements of the Waterloo Boy engine were retained on the new version.

The new design swept away the old steel-girder frame and the exposed final drive, and replaced them with unit construction and a fully enclosed transmission. The Model D was

Above, top: A spoked flywheel identifies this tractor as an early version of the John Deere Model D, the tractor that replaced the Waterloo Boy series.

Above: This sideways view shows the position of the twin-cylinder horizontal engine, probably the most famous engine series the tractor industry has produced.

shorter than its predecessor, and it was also much lighter, weighing 4,260lb (1,934kg) compared to the Model N's 6,200lb (2,812kg).

For John Deere two-cylinder enthusiasts the flywheel of the Model D is of particular interest. When production started the engine had a 26-in (66-cm) diameter spoked flywheel, but this was reduced to 24in (61cm) after the first 900 tractors had been built.

Flywheel

Another change came when the production total reached 5,755, and at this stage the spoked flywheel was replaced by a solid version. Model D tractors with a spoked flywheel, known to enthusiasts as "spoker Ds," are particularly prized by collectors because they date from the early production years.

Other changes introduced on the Model D included an increase in the cylinder bore to 6.75in (17.1cm) in 1928, and this plus other engine modifications introduced during the next few years raised the power output to 41.6hp (31kw) in 1935.

The original two-speed gearbox was replaced by three speeds in 1935, and the Model D was given new styling in 1939. Production continued in various versions until 1953.

Trail Blazer

The Model D was the first in a new line of rugged, reliable tractors that helped to establish John Deere as one of the most successful tractor companies in the United States, and it also contributed to the long-running success of the John Deere two-cylinder engine series.

Specifications

Manufacturer:	Deere & Co.
Location:	Waterloo, Iowa
Model:	D
Type:	General purpose
Power unit:	Two-cylinder horizontal engine
Power output:	30.4hp (22.5kw)
Transmission:	Two-speed gearbox
Weight:	4,260lb (1,934kg)
Production started:	1923

Below: Model D tractors in various versions and with steadily increasing power outputs were to remain in production for 30 years.

HART-PARR
🔧 **1924   Charles City, Iowa, USA**

HART-PARR 12-24E

By the early 1920s the Hart-Parr company was making serious efforts to move away from its heavyweight-tractor image and compete in the small and medium sectors of the tractor market, and the 12-24E model was one of a new generation of lightweight tractors. Introduced in 1924, it weighed 4,675lb (2,122kg) and was one of the smallest models in the Hart-Parr range.

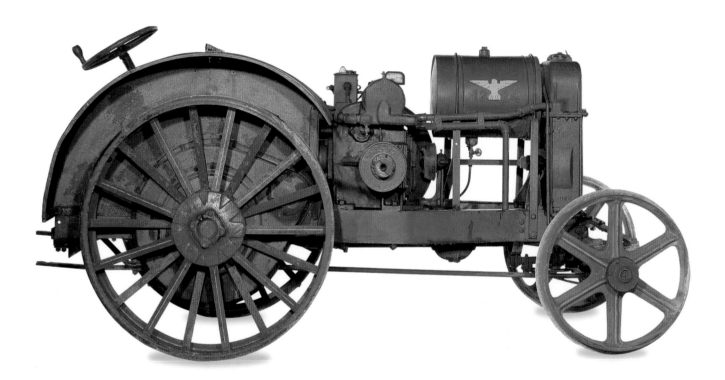

The power unit for the 12-24E was typical of a Hart-Parr design, based on a pair of horizontal cylinders, a valve-in-head layout, and an 800rpm rated speed. The bore and stroke measurements were 5.5in and 6.5in (13.9cm and 16.5cm), and the tractor's maximum output when the 12-24E was tested in Nebraska was almost 27hp (20kw).

Other Versions

The 12-24E was a development from the earlier 10-20 model, which was available in two different versions between 1921 and 1924. Both the 12-24 and the 10-20 shared basically the same engine, and the principal improvement to be found in the 12-24E model was a more up-to-date disk clutch to replace the previous model's band clutch.

The same engine, with an extra 0.25in (6mm) added to the cylinder bores, appeared once more in a new 12-24H tractor, available from 1928 as the replacement for the E.

The classic H version of the 12-24 was still in production when the Hart-Parr company lost its identity in the 1929 merger that formed the Oliver Farm Equipment Co., one of a series of mergers and takeovers that rationalized the tractor industry around that time.

Above: Hart-Parr's efforts to attract smaller-acreage customers included the little 12-24E lightweight tractor.

Specifications

Manufacturer: Hart-Parr Co.

Location: Charles City, Iowa

Model: 12-24E

Type: General purpose

Power unit: Two-cylinder horizontal engine

Power output: 27hp (20kw)

Transmission: Two-speed gearbox

Weight: 4,675lb (2,122kg)

Production started: 1924

HEIDER

✕ 1924   Rock Island, Illinois, USA

HEIDER 15-27

An Iowa farming family started the Heider Manufacturing Co., which began building tractors in 1911. The tractors were popular, and within a few years the Heiders sold their business to the Rock Island Plow Co. of Rock Island, Iowa.

Right: *The Heider name was retained after the Rock Island Plow Co. bought the tractor business from the Iowa farming family that had designed the tractors.*

Specifications

Manufacturer: Rock Island Plow Co.
Location: Rock Island, Illinois
Model: Heider 15-27
Type: General purpose
Power unit: Waukesha four-cylinder engine
Power output: 30hp (22.2kw)
Transmission: Friction drive with infinitely variable travel speeds
Weight: 6,290lb (2,856kg)
Production started: 1924

The new owners continued to base their tractors on the Heider design, and they also continued to use the Heider brand name and the distinctive trim along the sides of the canopy roof. They retained the Heider friction drive as well, another of the special design features. It provided an infinitely variable range of travel speeds, allowing the driver to match the engine and travel speeds more accurately to suit the working conditions and the equipment used.

Engine

When the Rock Island company introduced its new 15-27 tractor in 1924 it was still using both the Heider name and the existing friction drive. The Heider 15-27, like most Heider tractors, was equipped with a Waukesha four-cylinder engine. Cylinder measurements were 4.75-in (12-cm) bore and 6.75-in (17.1-cm) stroke, and the tractor's engine was positioned above the rear axle where the weight would assist the wheel grip.

The 15-27 remained in the Rock Island range for about three years and was one of the last tractors to carry the Heider brand name. The name was eventually dropped in the late 1920s, and the Rock Island Co. became part of the J. I. Case group in 1937.

HSCS

⚒ 1924 Budapest, Hungary

HSCS PROTOTYPE

Clayton and Shuttleworth was one of Britain's leading manufacturers of agricultural steam engines during the nineteenth and early twentieth centuries, and the big farms and estates in Hungary provided one of the company's best export markets.

Left: The original prototype tractor developed by the HSCS company still survives as an exhibit in the Hungarian Agricultural Museum.

Specifications

Manufacturer: Hofherr-Schrantz Clayton-Shuttleworth	
Location: Budapest, Hungary	
Model: Prototype	
Type: General purpose	
Power unit: Single-cylinder hot-bulb engine	
Power output: 15hp (11kw)	
Transmission: Single-speed	
Weight: N/A	
Production: 1924	

The switch from steam to tractor power caused serious problems for Clayton and Shuttleworth, and in 1912 it sold its Hungarian subsidiary. The buyer was a local farm-machinery manufacturer, and following the takeover the name was changed to Hofherr-Schrantz Clayton-Shuttleworth, or HSCS.

HSCS Tractors

The new company expanded, and by the early 1920s it had become Hungary's biggest manufacturer of stationary engines. It started its tractor development program in the early 1920s, and its first tractor powered by a semidiesel or hot-bulb engine was completed in 1924. It was a prototype model used for further development work, and it was powered by a single-cylinder two-stroke engine with 15-hp (11-kw) output, and it was capable of plowing between three and four acres at 7.5-in (19-cm) depth in a 10-hour day.

When the first HSCS production tractors arrived in 1925 they bore little resemblance to the prototype model. The single-speed transmission of the prototype was replaced by a three-speed gearbox, and the tractors' power output had increased to 20hp (14.9kw), but they still used a hot-bulb engine, which remained the standard power unit for HSCS tractors for more than 20 years.

CLETRAC
🛠 1925   Cleveland, Ohio, USA

CLETRAC K-20

It may, of course, have been just a coincidence, but the first model in the Cleveland Motor Co.'s Cletrac range of crawler tractors was known as the Model R, its second was identified by the letter H and tractor number three was the Model W—and the letters RHW are the initials of Rollin H. White, who was the principle shareholder in the Cleveland company.

Above: Cletrac's K-20 tractor arrived in 1925 to challenge the newly formed Caterpillar company in the crawler tractor market.

Right: Adding a power-takeoff shaft to the Cletrac K-20 model was a significant step forward in crawler tractor design.

Specifications

Manufacturer:	Cleveland Tractor Co.
Location:	Cleveland, Ohio
Model:	K-20
Type:	Tracklayer
Power unit:	Four-cylinder engine with overhead valves
Power output:	24.5hp (18kw)
Transmission:	Three-speed gearbox
Weight:	N/A
Production started:	1925

After that the lettering appears to be completely random, with the Model F followed by the K-20, then the 30A. The K-20 arrived in 1925, featuring the Cletrac controlled differential steering, and it was based on the earlier R, H, and W models and shared their compact overall size.

Design Features
The engine was a Cleveland built four-cylinder overhead valve design with 4in by 5.5in (10.1cm by 13.9cm) cylinders and a 24.5-hp

(18-kw) rated output, almost 5hp (3.7kw) more than the previous Model W. Possibly because of his previous automobile-industry background, Rollin White had a preference for high-speed engines for his tractors, and this was a feature of the K-20 engine with its 1,350rpm rated speed.

Principal improvements introduced on the K-20 were a belt pulley moved to the rear instead of the front of the tractor, and—a significant step forward in tracklayer design—it was also the first Cletrac to be equipped with a power takeoff.

RENAULT
✖ 1927 Billancourt, France

RENAULT PE

The PE model was introduced as a replacement for the H0 tractor, and it was also the first Renault model to break away from a design that was based on the original Renault light tank.

Another first for the PE was the addition of a manually operated implement-lift mechanism, and a push-button electric starter was available starting in 1933. The list of options also included solid rubber tires for the rear wheels, and it was also the first production tractor available with the new low-pressure inflatable tires introduced by Michelin in 1933.

Cooling System

Although the PE and the H0 look very different, one of the design features carried over from the previous model was locating the radiator behind the engine. On the PE model the radiator is upright instead of being slanted at an angle as on the previous tractors, and the vertical structure in front of the engine is an air cleaner with an enormous capacity.

The unconventional position of the radiator was not the only unusual feature of the PE cooling system. The PE's cooling fan was designed to form part of the clutch assembly in the base of the engine compartment, where it blew warm air out of the bottom of the

Above, top: In this photograph the side panels of the engine compartment have been removed, but their function was to force cooling air through the tractor's radiator.

Above: This Renault publicity shot shows a PE tractor equipped with the manually operated lift mechanism with a mounted plow.

compartment, drawing in cool air to replace it. Close-fitting steel panels enclosing both sides of the compartment ensured that most of the incoming air was drawn in through the radiator to cool the water.

The obvious advantage of placing the radiator at the rear of the engine is that it can allow a sloping hood line for improved forward visibility, as on previous Renault models, but this was not a feature of the PE.

The Renault design team may have realized that the cooling system was unnecessarily complicated because on later versions of the PE it scrapped the back-to-front arrangement, moving the radiator to the front of the engine, where it replaced the unnecessarily large air cleaner. In addition, a fan was placed behind the radiator to draw air directly through the radiator grille from front to back.

Disappointing Sales

Later developments of the PE model also included a narrow version for vineyard work, with the overall width reduced to 3ft 9in (1.14m). In spite of the options and design improvements, sales of the PE remained disappointing. This was partly due to the general economic situation at the time and partly because of intense competition from American imports and from the French-built version of the British Austin tractor. PE sales totaled 1,771 between 1927 and the end of production in 1936, amounting to a modest average of fewer than 200 per year.

Specifications

Manufacturer: Renault	**Power output:** 20hp (14.8kw)
Location: Billancourt, France	**Transmission:** Three-speed gearbox
Model: PE	**Weight:** 3,960lb (1,800kg)
Type: General purpose	**Production started:** 1927
Power unit: Four-cylinder 127-cubic-inch (2.1-liter) engine	

Left: *The large rectangular structure at the front of the PE's engine compartment is an enormous air cleaner for the engine.*

FATE-ROOT-HEATH

✖ 1933 Plymouth, Ohio, USA

SILVER KING

The Fate-Root-Heath company of Plymouth, Ohio, chose a lightweight model when it moved into the tractor market in 1933. It was the company's first tractor venture and it was named the Silver King. The power unit chosen was the popular Hercules IXA, a four-cylinder engine with a rated output of 20hp (14.8kw).

Left: *The Silver King offered many small-acreage farms an opportunity to make the change from horses or mules to tractor power.*

Specifications

Manufacturer: Fate-Root-Heath Co.

Location: Plymouth, Ohio

Model: Silver King

Type: General purpose

Power unit: Hercules four-cylinder gasoline engine

Power output: 20hp (14.8kw)

Transmission: Four-speed gearbox

Weight: N/A

Production started: 1933

Many of the companies moving into tractor production for the first time misjudge the market and encounter serious problems, but the new Silver King tractor was a first-time success. There was a growing demand in the early 1930s for small, economical models to suit the thousands of farmers buying their first tractor to replace horses or mules, and the Silver King was a popular choice at a time when some of the leading manufacturers had still not developed their own lightweight tractors.

Tricycle Version

Silver King production started with the standard four-wheel model, and it was among the first tractors offered with the recently developed rubber tires.

A slightly bigger engine was fitted to the 1936 model, and at the same time a new tricycle version was introduced. The three-wheeler was tested in Nebraska, where it developed 19.74hp (14.7kw) in the brake test and achieved a top speed of 25mph (40km/h) in fourth gear.

The company's sales were hit by increasing competition as more manufacturers moved into the lightweight end of the market, and new models announced by Silver King in 1940 failed to recapture the company's earlier success. Production ended in the 1950s.

ALLIS-CHALMERS
1933 Milwaukee, Wisconsin, USA

ALLIS-CHALMERS WC

The WC model was introduced by Allis-Chalmers to capture a share of the highly important rowcrop-tractor market in the United States and Canada, and it was a big success. It was announced in 1933, and the sales total had passed the 170,000 mark when production ended in 1948.

Above: *This front view shows the offset steering position, the dual front wheels, and the high clearance under the rear axle.*

Right: *The WC's profile shared some similarities with Allis-Chalmers Model B rowcrop tractor.*

Specifications

Manufacturer: Allis-Chalmers Manufacturing
Location: Milwaukee, Wisconsin
Model: WC
Type: Rowcrop
Power unit: Allis-Chalmers four-cylinder engine
Power output: 21hp (15.5kw)
Transmission: Four-speed gearbox
Weight: 3,190lb (1,448kg)
Production started: 1933

Allis-Chalmers also offered a WF version with a full-width front axle to give a standard wheel layout, but the WC rowcrop model outsold the WF by a 20 to 1 ratio, providing an example of how important the rowcrop-tractor market is in North America.

Two-Plow Tractor

The WC model was described as a two-plow tractor—designed to turn two furrows—and it was powered by an Allis-Chalmers–built four-cylinder engine. It was an unusual design because the piston bore and stroke were 4.0in (10.1cm), and an engine with this feature is sometimes described, misleadingly, as "square." The rated power output was just under 21hp (15.5kw), and when the WC was tested in Nebraska it reached 21.48hp (16.2kw) in the maximum-load test.

When the WC made its first appearance the front-end styling was flat, but in 1938 the Allis-Chalmers stylists introduced a more up-to-date streamlined appearance with a rounded grille in front of the radiator, and the tractor in the photograph here was built after the new look was introduced.

CASE

⚒ 1936 Racine, Wisconsin, USA

CASE MODEL R

The Case company was quick to appreciate the importance of good styling, and the Model R is a fine example. The appearance is neat and well balanced, and the design of the radiator grille is distinctive and eye-catching, having been inspired by the shape of an ear of wheat, according to one theory.

Above: *The distinctive pattern on the Model R radiator grille is said to have been based on the shape of an ear of wheat.*

Specifications

Manufacturer: J. I. Case
Location: Racine, Wisconsin
Model: R
Type: General purpose
Power unit: Waukesha L-head engine
Power output: 20hp (14.8kw)
Transmission: Three-speed gearbox
Weight: 4,140lb (1,880kg)
Production started: 1936

Later versions of the Model R also benefit from the new Flambeau Red paint color introduced by Case in 1939 to make its tractor range more eye-catching.

Case announced the Model R in 1936 to fill a gap at the lower end of the power scale. Lacking a suitable power unit of its own, Case's choice for the Model R was a gasoline engine supplied by Waukesha and developing about 20hp (14.8kw) at 1,400rpm. The power was delivered through a three-speed gearbox. Rubber tires were included on the options list for the Model R.

Versions

Three versions of the Model R were available. These were the standard-tread Model R, the RC with a tricycle-wheel layout, and an industrial model identified as the RI. The tractors were introduced at a time when the Case company was beginning to recover from financial difficulties and management problems, and the little R series tractors played their part in the recovery.

Production of the Model R continued until 1940, when it was phased out to make way for the new V-series tractors.

Above, left: *A standard-tread version of the Case Model R finished in the Flambeau Red paint finish adopted by Case in 1939 to give its tractors more visual appeal.*

RANSOMES

1936 Ipswich, Suffolk, England

RANSOMES MG TRACKLAYERS

Ransomes was one of the leading manufacturers of agricultural steam engines during the nineteenth century, but its most notable success in the tractor market was on a much more modest scale.

Right: *Providing basic tractor power on a small scale, the little Ransomes MG crawler tractor was available in various versions for 30 years.*

Above: *The Ransomes decal on the front of the fuel tank of its MG series "motor cultivator," or market-garden tractor.*

Specifications

Manufacturer: Ransomes, Sims and Jefferies
Location: Ipswich, Suffolk, England
Model: MG
Type: Tracklayer
Power unit: Sturmey-Archer single-cylinder engine
Power output: 4.25hp (3.3kw)
Transmission: Single speed forward and reverse
Weight: N/A
Production started: 1936

The Ransomes MG was a mini-sized crawler tractor designed mainly for market gardens. The first production version, called the MG2, was available in 1936 powered by a 37-ci (600-cc) single-cylinder Sturmey-Archer engine. The design included a centrifugal clutch delivering the power through reduction gears, and a pair of crown wheels, one for forward travel and the other for reverse.

Tracks

The tracks were a Roadless design with rubber joints, and the width between track centers could be adjusted. For rowcrop work a set of crop dividers to push plants away from the tracks and minimize damage was on the options list, and Ransomes also produced a wide range of special implements to suit the small size and modest power of the MG.

MG2 production continued until 1949 when the MG5 version arrived. The new model was powered by a 37-ci (600-cc) Ransomes side-valve engine producing up to 4hp (2.98kw), and this was followed by the diesel-powered MG40. Approximately 15,000 MG tractors were built during the 30-year production run, and, although most of them were used in small-scale crop production, others were used for a varied range of jobs including harvesting salt in East Africa, and some were exported to Holland where their small size allowed them to be ferried across drainage dykes on small boats.

ALLIS-CHALMERS
✖ 1937 Milwaukee, Wisconsin, USA

ALLIS-CHALMERS MODEL B

The Allis-Chalmers entry in the late 1930s competition for small tractor sales was the popular Model B. It was a big success, with sales totaling more than 127,000 tractors during a 20-year production run starting in 1937, and the total included some Model B tractors built in England from 1947.

Some of the Model B's success was due to its versatile design. It was a popular general-purpose tractor on livestock farms, with a turning circle of 7ft 8in (2.3m) to provide good maneuverability, but it was also designed for rowcrop work, and this was an important factor in its sales success.

The design included plenty of underside clearance to provide space for mid-mounted equipment such as inter-row hoes, and the distinctive slimline shape allowed the driver a good view for accurately steering between the crop rows. There was also a generous amount of wheel track adjustment with settings available from 40in to 52in (102cm to 132cm) to suit crops grown on different row widths.

Engine
Allis-Chalmers built a four-cylinder engine for the Model B, with a bore and stroke of 3.25in

Above, top: The Allis-Chalmers B starting handle was mounted under the rear of the fuel tank, where it was easily accessible when needed.

Above: With its neat styling and good design features for rowcrop work, the Allis-Chalmers Model B was popular throughout the 1940s.

by 3.5in (8.3cm by 8.9cm) for tractors built before 1944. Engine capacity was increased slightly by using a wider cylinder bore for tractors built from 1944 until production of the Model B ended in 1957. Power output for the smaller engine was 14hp (10kw) at the 1,400rpm rated engine speed, but this was increased to 16hp (12kw) for the later version.

A three-speed gearbox provided a maximum travel speed of 7.75mph (12.5km/h) at the rated engine rpm, and driver comfort on later versions was improved by providing a fully padded seat cushion and backrest instead of the basic metal-pan seat mounted on a simple spring included in the earlier specification. As well as offering more power, later versions of the Model B were also equipped with additional features such as a hydraulically operated implement lift instead of the standard mechanical version, and electric lights were added to the options list.

UK Production

Model B popularity extended to British farms, where it sold so well that the Allis-Chalmers company joined other leading North American tractor manufacturers by setting up a UK assembly plant. Production started in 1947, and a diesel version powered by a Perkins P3 engine was available in the UK starting in 1953.

In addition to the diesel option, UK models were available with the choice of steel wheels or rubber tires, and the options list also included a three-speed belt pulley and a power-takeoff kit.

Specifications

Manufacturer: Allis-Chalmers
Location: Milwaukee, Wisconsin
Model: B
Type: Rowcrop
Power unit: Allis-Chalmers four-cylinder engine
Power output: 14hp (10kw) (early version)
Transmission: Three-speed gearbox

Weight: 2,620lb (1,189kg)
Production started: 1937

Below: An important feature of the Model B design was the space beneath the engine and transmission for carrying mid-mounted hoes and other inter-row equipment.

FORD

1939 **Dearborn, Michigan, USA**

FORD MODEL 9N

As the cracks began to show in the business relationship between Harry Ferguson and David Brown, Ferguson set sail for America to seek a new partnership, and the person he had in mind was Henry Ford.

It was a shrewd move. The Ferguson-Brown partnership had launched Ferguson's three-point linkage system and proved its significant advantages, but commercial progress was slow and Ferguson needed much bigger resources to achieve the international impact that he needed for his equipment.

Ferguson and Ford

Nobody in the 1930s tractor industry had bigger resources than Henry Ford, and he was actively looking for a new tractor design to replace the aging Model N Fordson from England. Henry Ford had spent many hours designing experimental tractors in the workshop at Fair Lane, the mansion he had built in Dearborn. However, when Harry Ferguson arrived in 1938, Ford was still looking for the right idea.

The meeting, held in the grounds of Fair Lane, had been arranged to give Ford an opportunity to see the Ferguson System in

Above: *The Moto-Tug was a special version of the 9N designed for working on wartime airfields, but later used as a small general-purpose industrial tractor.*

action. Harry Ferguson had brought one of his Model A tractors and some implements from England, and Ford arranged for a Fordson Model N and an Allis-Chalmers Model B to be brought from his own farm for comparison.

The Model A easily outperformed its rivals, and Henry Ford, a farmer's son, quickly appreciated the advantages. By the end of the demonstration he and Ferguson had agreed to form a partnership to design, build, and market a new Ferguson System tractor, with Ford providing much of the finance and the production facilities; Ferguson's responsibility was to set up a dealer network to market the tractors in North America.

New Tractors

Henry Ford's checkbook ensured that the project was completed quickly and efficiently. The new tractor, called the Ford 9N or Ford

Tractor with Ferguson System, was ready for production only eight months from the original demonstration, and it was launched at a lavish party held on the Ford farm for 500 VIP guests.

The new tractor was powered by a Ford L-head engine, with the power delivered through a three-speed gearbox with a top speed of 6mph (9.6km/h) on the road. Rubber tires were standard equipment. The list of options included steel wheels, road lights, and a belt pulley. It was suggested—but perhaps not seriously—that as the engine was so quiet the 9N tractor might be equipped with a built-in radio.

Specifications

Manufacturer: Ford Motor Co.	**Transmission:** Three-speed gearbox
Location: Dearborn, Michigan	**Weight:** 3,375lb (1,532kg)
Model: 9N	**Production started:** 1939
Type: General purpose	
Power unit: Ford four-cylinder engine	
Power output: 23hp (17kw) (maximum)	

Below: *When Harry Ferguson and Henry Ford formed a partnership to develop a new tractor, the result was the Ford 9N, the most advanced small tractor of its day.*

CASE

1939 **Racine, Wisconsin, USA**

CASE LA

When Case introduced the new LA tractor in 1939 it was an important event for the company. The tractor it replaced, the Case L, had built up an excellent reputation as one of the best tractors in its power range, and the design team for the new LA model must have realized it was a hard act to follow.

To some extent the Case designers played safe and simply retained most of the mechanism of the L. The components that were carried over into the new model included the four-cylinder engine, a valve-in-head design with 400.6-ci (6.6-liter) capacity. When the engine first appeared in the L model the output had been about 40hp (29.8kw), but this was increased to almost 50hp (37.2kw) for the new LA tractor.

Transmission and Drive

The transmission was also based on the L model and was unconventional by 1940s standards. The clutch was controlled by a hand-operated lever instead of a foot pedal, and when the lever was moved into the "drive" position it forced a pair of metal plates together, sandwiching the clutch disk between them. Unlike a conventional clutch, the over-center action of the lever held the plates together without using compression

Above: *The four-cylinder engine for the Case LA tractor produced up to 48hp (35.5kw) and was an uprated version of the power unit in the previous Model L.*

springs, but there was a set of springs to push the plates away from the disk when the drive was disengaged.

The final drive was also unusual. It consisted of a pair of chains and sprockets to transfer the power from the gearbox to the rear axle. Although this was old-fashioned by the time the LA tractor was launched, it was also a simple, reliable arrangement that was easy to maintain and was already familiar to many farmers.

Styling

Although, apart from a few detailed improvements, the L and LA tractors shared basically the same mechanical features, and it was the styling that provided the obvious differences. The late 1930s were a time when the tractor stylists, particularly in the United States, were adopting brighter, more striking paint colors and borrowing rounded bodywork from the car industry, and the LA was an early example of the new look adopted by the Case company, including a bright orange color called "Flambeau Red."

Styling is obviously a matter of taste, and the rounded shape of the LA certainly attracted customers and continued to be fashionable throughout the 1940s. There are also those who think that the stylists who were responsible for the simple, clean lines of the old L tractor had produced one of the classic designs of the 1930s, and the shape of the LA was, perhaps, not an improvement.

Above: *An unusual design feature on the big LA tractor was the hand-operated lever to control the clutch.*

Specifications

Manufacturer: J. I. Case	**Power output:** 48hp (35.5kw)
Location: Racine, Wisconsin	**Transmission:** Four-speed gearbox
Model: LA	**Weight:** 5,940lb (2,700kg)
Type: General purpose	**Production started:** 1939
Power unit: Case four-cylinder engine	

Below: *This cutaway diagram shows the final drive of the LA tractor, using a chain and sprockets to deliver the power to the rear axle.*

OPPERMAN
1946 Borehamwood, Hertfordshire, England

OPPERMAN MOTOCART

The Motocart was designed in 1945 by a farmer. He decided that a small rough-terrain load-carrying vehicle would outperform the horse and cart he was using for moving loads on the farm and for local transportation work.

A prototype built in the farm workshop was so successful that he took the idea to S. E. Opperman, an engineering company that was making special strakes or grips to fit over tractor tires for improved traction in difficult conditions. Opperman improved the basic design of the load carrier, called it the Motocart and started building it in 1946.

Design

The design was ingenious, with a single large-diameter wheel at the front to provide traction and for steering the vehicle, plus a small air-cooled engine actually attached to the offside of the front wheel. The small wheels at the rear were equipped with drum brakes, and the driver was positioned near the front of the vehicle, between the front wheel and the rear load compartment.

A single-cylinder four-stroke engine provided about 8hp (6kw) to drive the front wheel. The power was delivered by a chain and sprocket to a single-plate clutch and a compact four-speed gearbox providing a top speed of 11mph (17.7km/h) on the road.

The Motocart's load capacity was 1.5 tons (1.52 tonnes) carried in a choice of fixed and tipping bodies, and the work rate for on-farm transportation work was said to be three times faster than that achievable by a horse and cart.

Above: *The initial idea of a transportation tractor for farms was adopted by the Opperman company in the Motocart.*

Specifications

Manufacturer: S. E. Opperman	
Location: Borehamwood, Hertfordshire, England	
Model: Motocart	
Type: Transportation tractor	
Power unit: Single-cylinder air-cooled engine	
Power output: 8hp (6kw)	
Transmission: Four-speed gearbox	
Weight: 3,300lb (1,500kg)	
Production started: 1946	

BEAN
✖ 1946   Brough, Yorkshire, England

BEAN TOOL CARRIER

Not surprisingly, the Bean tool carrier was designed by a Mr. Bean, a vegetable grower from Yorkshire, England. His first machine was built for his own use in about 1945, and when this attracted interest from other growers he made more of the tool carriers to sell locally.

Below: *A Yorkshire vegetable grower designed the Bean 8-hp (6-kw) tool carrier and a range of special implements for rowcrop work.*

Specifications

Manufacturer: Humberside Agricultural Products
Location: Brough, Yorkshire, England
Model: Bean
Type: Tool carrier
Power unit: Ford four-cylinder engine
Power output: 8hp (6kw)
Transmission: Three-speed gearbox
Weight: 1,344lb (610kg)
Production started: 1946

As the demand continued to increase Mr. Bean arranged for Humberside Agricultural Products to take over the production in 1946, and it continued to build it for about 10 years.

Two Versions
The tool carrier was available in three- and four-wheel versions, but the three-wheeler was the most popular model. Both versions were based on a rectangular steel frame with two large driving wheels at the rear, and the tricycle version had a single wheel at the front with tiller steering. An 8-hp (6-kw) Ford industrial engine mounted over the rear wheels provided

the power, and this was linked to a three-speed Ford gearbox. The driver's seat was in front of the engine, providing an almost uninterrupted view of the ground in front of the tractor and of mid-mounted implements, and equipment such as hoes or light harrows could also be attached to the rear of the frame.

Bean tool carriers were used mainly for working in field-scale vegetable crops, but they were also popular for controlling weeds in sugar beet. As well as inter-row cultivators and hoes, the attachments list for the Bean included a sprayer, a six-row drill for sowing vegetable seeds, and a fertilizer distributor.

FERGUSON
🔧 **1946 Coventry, Warwickshire, England**

FERGUSON TE-20

When the American-built Ford 9N tractor with its Ferguson System hydraulics and rear linkage proved to be an outstanding success, Harry Ferguson expected that Ford would extend production to its factory at Dagenham in England.

Exactly why this did not happen is not clear. It may have been due to the practical problems of changing production during the war years, but another suggestion is that the Ford directors at Dagenham were reluctant to involve Ferguson who, rightly or wrongly, had a reputation for being difficult to work with. Instead, when World War II ended they replaced the Model N tractor with the Fordson E27N Major.

Ferguson and Standard

Harry Ferguson realized that he would have to make his own arrangements to build a Ferguson System tractor in England, and he subsequently signed an agreement with the Standard Motor Company, a major car manufacturer. Standard was a good choice because it had plenty of space available at its factory in Banner Lane, Coventry, and it also had a good reputation for quality. The arrangement was somewhat similar

Above, top: The Ferguson System three-point linkage on the back of a TE series tractor from the Banner Lane factory.

Above: Ferguson tractors were built at the Banner Lane factory near Coventry under an agreement with the Standard Motor Co., which was responsible for production.

to Ferguson's previous agreements with David Brown and Henry Ford. Ferguson controlled the design and marketing, while his partner provided the factory and looked after the tractor's production.

Production

Small-scale production began at the end of 1946. The tractor was the Ferguson TE-20 and it closely resembled the American-built Ford 9N tractor, with both sharing the same battleship-gray paint finish and basically similar styling. The major differences included a four-speed gearbox in the new tractor, and the engine for the first two years of TE production was supplied by Continental and featured overhead valves, 1,179.6-ci (1,966-cc) capacity, and a maximum output of 24 hp (17.8 kw).

The replacement for the Continental power unit was an engine of approximately similar size and power built by the Standard company, and this engine was also adopted to power the Standard Vanguard car, pickup truck, and van.

When a diesel model was added to the TE series in 1951, Ferguson chose a new engine designed by Standard, and the tractor was called the TE-F20.

Developments

Other TE series developments included an agreement to build Ferguson tractors in France, where the locally built Ferguson rapidly overtook the Renault as the top-selling tractor in France. The biggest overseas project was in the USA, where Ford's decision to stop supplying tractors to Harry Ferguson's North American marketing company left him with a big dealer network and no tractors to sell. Ferguson was able to supply TE series tractors from England while he set up new production facilities in Detroit.

The American-built Ferguson was called the TO-20. Although it was based on the TE design, the TO-20 was equipped with a Continental engine and a number of other American-sourced components.

Specifications

Manufacturer: Standard Motor Co.

Location: Coventry, Warwickshire, England

Model: Ferguson TE-20

Type: General purpose

Power unit: Four-cylinder engine

Power output: 24hp (17.8kw) (maximum)

Transmission: Four-speed gearbox

Weight: 2,460lb (1,117kg)

Production started: 1946

Below: *This was the first TE series tractor to be built in the Banner Lane fatory. It was used as a light factory vehicle, but was later restored and moved into the museum at the factory.*

FORD

1947 Dearborn, Michigan, USA

FORD 8N

Henry Ford continued to control the fortunes of the huge company he had built until 1945 when, at the age of 82, he handed over the reins to his grandson, Henry Ford II. As Henry Ford senior's influence in the company declined, his partnership with Harry Ferguson also entered its final phase.

In 1945 the Ford organization was losing money, and harsh decisions were necessary to halt the rapidly mounting losses. One of the decisions was to end the "handshake agreement" between Henry Ford senior and Harry Ferguson. The agreement had produced the 9N and 2N tractors, built by Ford and incorporating Ferguson System hydraulics, but marketed by a company controlled by Ferguson.

Legal Action

Ford's American tractor operation was one of the problem areas identified in the 1945–46 financial review, and at the end of 1946 Henry Ford II announced that Ford would stop supplying tractors to the Ferguson marketing company after a six-month notice period, and that the 9N/2N tractors would be replaced by a new model.

This later prompted legal action by Harry Ferguson, who sued the Ford company for patent infringements and damages in an action involving 200 lawyers and more than a million documents. The outcome favored Ferguson, but he was awarded less than $10 million, instead of the $340 million he had claimed.

Above, top: *The new 8N tractor bore a superficial resemblance to the old 9N model, and it also included the full Ferguson System linkage and hydraulics.*

Above: *The Ford 8N's launch brought the Ford–Ferguson partnership to an acrimonious end in what was an extremely bitter and expensive legal action.*

Another result of the legal action was an agreement to make design changes to the new Ford tractor in order to stop infringement of Ferguson's patents. The new tractor was the Ford 8N, launched in July 1947 to replace the 9N/2N and offering a list of more than 20 design improvements.

The 8N started its commercial life with the full Ferguson System rear linkage and hydraulics—later modified following the court decision—but the Ferguson-approved gray paint finish was replaced by a pale gray and red, and the three-speed gearbox specified by Harry Ferguson for the previous model was replaced by a four-speed version in the new tractor.

Success

In spite of the legal action and the modifications, the 8N was one of the biggest success stories

the tractor industry has seen, with production peaking at more than 100,000 tractors per year in 1948 and 1949. The only other tractor that has broken the 100,000 per year barrier was Henry Ford's Fordson Model F. However, Harry Ferguson was able to achieve the same distinction by combining the production totals for his British-built TE series tractors and their American TO equivalent.

Ford 8N production continued until 1953 when the new NAA tractor arrived as part of the Ford company's Golden Jubilee celebrations.

Specifications

Manufacturer: Ford Motor Co.
Location: Dearborn, Michigan
Model: 8N
Type: General purpose
Power unit: Four-cylinder engine
Power output: 21hp (15.5kw) (maximum)

Transmission: Four-speed gearbox
Weight: 2,710lb (1,230kg)
Production started: 1947

Left: *This publicity photograph from the Ford archives shows Ford 8N tractors at the end of the 8N production lines at the Ford factory in Dearborn, Michigan.*

URSUS

🔧 **1947   Ursus, Warsaw, Poland**

URSUS C45

Ursus tractors first appeared in the early 1920s, and the first model was a copy of the International Harvester Titan 10-20 tractor built under a license agreement. Tractor production was one of the priorities in the postwar reconstruction of Polish industry, and the Ursus factory started building tractors again in 1947.

Above: *A fully cushioned seat mounted on a coil spring was part of the package of driver comfort features on the Polish-built Ursus C45.*

The 1947 tractor was the C45 and, as the photographs show, it was a Lanz Bulldog lookalike—in fact the two tractors are so similar that Bulldog replacement parts fit the Ursus.

Hot-Bulb Engine

The C45 was the result of another license agreement, and the aging Bulldog design became a familiar sight and sound on Polish farms. It featured the familiar Lanz hot-bulb or semidiesel engine, and the standard version featured the usual blowlamp to heat the cylinder head, while the wood-rimmed steering wheel doubled as a starting handle to turn the tractor's engine over.

Customers who were reluctant to spend 20 minutes or so to start the engine from cold could specify an optional electric starter. This system used a trembler coil to provide the spark to ignite gasoline, and the engine continued to

Above, left: *This picture of the Ursus C45 semidiesel-powered tractor shows how closely the design resembles a Lanz tractor.*

run on gasoline until the cylinder head was sufficiently hot to continue running on diesel or paraffin.

The gasoline/electric system speeded up the starting process, but it did not overcome the quite strong possibility that the engine would run backward. This is a characteristic of semidiesels, and in the case of the C45 it would mean that the four forward speeds and one reverse are converted into four reverse speeds and one forward.

Driver Comfort

There are just a few indications on the C45 that driver comfort and convenience were beginning to emerge as significant design features. The seat on the C45 tractor is on a coil spring to allow vertical movement, and it is cushioned—although the cushions in the photograph at left are not the originals. The driving position is also offset, which allows better forward visibility, but also leaves more space for climbing aboard, and the most surprising comfort feature is the leaf spring suspension under the front axle to give the driver a smoother ride.

Production of the Ursus C45 was to continue well into the 1950s, when more modern designs took over; however, the policy of relying on imported technology that had started with the Titan and the Bulldog remained in place. There were close links between Ursus and the Zetor tractor company in Czechoslovakia, and during the 1980s Massey Ferguson provided the know-how to build a new Ursus factory where some MF models and Perkins engines were built under license.

Below: A massive transversely mounted leaf spring under the front axle of the Ursus C45 provided one contribution to driver comfort.

Specifications

Manufacturer: Ursus	**Transmission:** Four-speed gearbox
Location: Ursus, Warsaw, Poland	**Weight:** N/A
Model: C45	**Production started:** 1947
Type: General purpose	
Power unit: Semidiesel engine	
Power output: 45hp (33.3kw)	

FORDSON
⚒ 1951   Dagenham, Essex, England

NEW FORDSON MAJOR

In spite of its sales success the Fordson E27N had been introduced as a makeshift model to maintain sales while the Ford engineers were working on the New Fordson Major.

As well as the new tractor, they were also developing a completely new engine. The new engine was a diesel, and work had started on the project at the Ford factory at Dagenham in 1944 in spite of considerable opposition from some of the company's senior management. The opponents pointed out, quite correctly, that previous efforts to develop diesel-powered wheeled tractors had achieved little success, and there was no evidence that farmers would be willing to switch to the new fuel.

In spite of the doubts the development program continued, and the new tractor was launched at the 1951 Smithfield Show in London, with deliveries starting the following year. Customers were offered three engine options using gasoline, gasoline/paraffin, and diesel fuel. The diesel version had a 218.5-cubic -inch (3.6-liter) capacity with a 16:1 compression ratio, the paraffin burner was also 218.5 cubic inches but with 4.35:1 compression ratio, and the gasoline engine capacity was

Above: *The new Fordson Major was launched in 1951, with deliveries starting the following year. It was followed by the Super Major and Power Major versions.*

197.8 cubic inches (3.26 liters) with a 5.5:1 ratio. Power output was initially rated at 40hp (29.6kw) at 1,700rpm, but this was increased on later versions of the tractor.

Diesel Success

The diesel engine became a big success. It was one of small group of British engines that helped to introduce a new generation of diesel power with high-speed, multicylinder units that were smoother than previous diesels and easier to start. By the end of the production run of the New Major and its Super Major and Power Major derivatives more than 90 percent of UK customers were specifying the diesel version.

It was not just the engines that pushed the New Major up the UK sales charts and won export success as well. This was the first complete break from the original Fordson

Model F design of 1917, and it was more advanced technically and more capable than previous models in every respect. It combined up-to-date styling with a technical specification that included a smooth, dual-ratio gearbox with six forward speeds. It was also well built, and the New Major soon earned a reputation for reliability and durability.

The New Major was also popular as the engine and transmission for the growing number of four-wheel drive and crawler specialists, and the Fordson provided the power for tractors from County, Doe, and Roadless.

Above: *The Fordson Power Major was available from 1958 and based on the New Major, but with a number of detailed improvements.*

Specifications

Manufacturer: Ford Motor Co.	**Transmission:** Six-speed gearbox
Location: Dagenham, Essex, England	**Weight:** 5,308lb (2,409kg)
Model: New Fordson Major	**Production started:** 1951
Type: General purpose	
Power unit: Ford four-cylinder diesel engine	
Power output: 40hp (29.6kw)	

Below: *Toward the end of the Fordson Major production run more than 90 percent of UK customers were specifying the diesel version.*

PLATYPUS

1952 **Basildon, Essex, England**

HOWARD PLATYPUS 30

Arthur "Cliff" Howard was born in Australia, and in the late 1930s he moved to the UK to set up the Howard Rotavator company to build the tractor-powered rotary cultivator he had designed. The company was highly successful, at one stage claiming to be the world's biggest farm machinery manufacturer.

Above: *Various versions of the Platypus tractor were available during the six-year production run, all powered by Perkins diesel engines.*

Specifications

Manufacturer: Platypus Tractor Co.
Location: Basildon, Essex, England
Model: Howard Platypus 30
Type: Tracklayer
Power unit: Perkins P4 diesel engine
Power output: 34hp (25.2kw)
Transmission: Six-speed dual-range gearbox
Weight: 5,684lb (2,578kg)
Production started: 1952

Above, left: *Platypus Tractor Co. was a subsidiary of the Howard machinery company, and its crawler tractors were built at a factory in Basildon, Essex, in England.*

The first Platypus crawler tractors arrived in about 1950, powered by a Standard gasoline engine, and their success encouraged the Howard company to open a new factory at Basildon, Essex, in England, specifically for tractor production, starting in 1952.

Most of the Basildon-built tractors were equipped with Perkins engines, and the most popular model was the Platypus 30 with a P4 engine developing 34hp (25.2kw). There was also a 71-hp (52.9-kw) version powered by a Perkins R6 engine.

Presumably it was the Howard family's Australian connections that prompted the Platypus name for the tractors. A platypus is an Australian marsupial, and according to the Howard sales leaflet for the tractor, it is small but powerful, hard working, and as much at home on the water as on dry land.

Disappointing Sales

In spite of the Platypus name and development of a Bogmaster version with extra-wide tracks designed for working in extremely soft ground conditions, sales volumes for the tractor were disappointing, and production of the Platypus ended in 1958 when the decision was taken to close the loss-making Basildon factory.

MAN
⚒ **1952 Nürnberg, Germany**

MAN AS 440A

Not surprisingly the Maschinenfabrik Augsburg-Nürnberg company abbreviated its name to MAN, and this became a familiar brand name in the German tractor market during the 1950s and 1960s.

Right: *In a less safety-conscious age many European tractor manufacturers added a passenger seat to one of the fenders, or mudguards, of their tractors.*

Specifications

Manufacturer: MAN Ag

Location: Nürnberg, Germany

Model: AS 440A

Type: General purpose

Power unit: MAN four-cylinder diesel engine

Power output: 40hp (29.6kw)

Transmission: Six-speed gearbox

Weight: 4,675lb (2,120kg)

Production started: 1952

Germany has a long tradition of diesel-engine development, and MAN is one of the leading manufacturers. When the AS 440A tractor was introduced in 1952 it was powered by a D9214 series four-cylinder diesel engine built at the MAN factory in Nürnberg. Most of the German-built tractors during the 1950s were relatively low horsepower models designed for small-acreage family farms, many of them run on a part-time basis, but the AS440A was one of the exceptions.

It was a big tractor by contemporary standards, with the engine delivering a maximum output of 40hp (29.6kw) at 2,000rpm, and the tractor in working trim with a full fuel tank weighed just over two tons.

Versions

Two- and four-wheel drive versions were available, and the specification included a six-speed gearbox with almost 17mph (27km/h) available in top gear. MAN tractors are popular with German vintage-tractor enthusiasts, and the well restored model in the photograph includes the optional cab and road lights. The passenger seat on the left-hand fender, or mudguard, was a popular feature on many tractors used in Europe during the 1950s.

JOHN DEERE

✗ 1956 Waterloo, Iowa, USA

JOHN DEERE 820

The most immediate difference between John Deere's new 20 series tractors and the models they replaced was the distinctive two-tone green-and-yellow paint finish; however, there were also some significant performance and driver-comfort improvements.

The first two 20 series tractors, the 320 and 420 models from the John Deere factory in Dubuque, were announced in 1955, with the remaining models following in 1956. There were basically six models in the series, starting with the little one-plow 320 and including the 820 diesel at the top of the range.

Performance

A vertical gasoline engine with cylinders of 4in by 4in (10.1cm by 10.1cm) powered the 320, which was built in both standard and utility versions. The 420, with 27-hp (20.1-kw) output in its Nebraska tests, was the most versatile 20 series model, available in standard, utility, rowcrop, high-clearance, tricycle, and low-profile wheeled versions, plus a tracklayer with the choice of two different track options.

There were single-wheel, twin-wheel, and wide-axle versions of the 520, the 620 was built as a sleek-looking orchard model and could also be equipped for burning liquefied petroleum

Above: *The top model in John Deere's new 20 series tractors was the diesel-powered 820, with the power output boosted to 64hp (47.7kw) during the two-year production run.*

gas, and there were hi-crop, twin-wheel, single-wheel, and wide-axle options for the 720.

The diesel-powered 820, available as a standard model only, started its production life with the same twin-cylinder engine that had powered the previous 80 model, and the small gasoline-fueled starter motor was also carried over from the 80. Engine updates during the two-year production run boosted the 820's power rated output to 64hp (47.7kw).

Driver Comfort

Evidence that driver comfort was moving up the priority list included the new Float-Ride seat featuring a rubber torsion-spring suspension with an adjustment to suit different driver weights.

There was also a hydraulic shock absorber under the seat, the cushion and backrest were of foam rubber, and cushioned armrests were an option. Foot space on the driver's platform was increased, and the instruments and controls were reorganized to make them easier to use.

Design improvements to the hydraulics to make the three-point linkage system more accurate and give some degree of weight transfer were called the Custom Powr-Trol system, and the technical developments on the engines included modifications to the cylinder head and piston design to increase combustion chamber turbulence and boost power output and fuel efficiency.

Above: *The big twin-cylinder diesel engine on the 820 tractor was started by a small gasoline engine, which meant carrying a second fuel supply.*

Specifications

Manufacturer: Deere & Co.

Location: Waterloo, Iowa

Model: John Deere 820

Type: General purpose

Power Unit: Deere two-cylinder horizontal diesel engine

Power output: 72.8hp (59kw) (maximum)

Transmission: Six-speed gearbox

Weight: 8,729lb (3,963kw)

Production started: 1956

Below: *Smallest of the 20 series tractors with its new two-tone paint finish was the Dubuque-built 320 model introduced in 1955.*

LANZ
✂ 1956 Mannheim, Germany

LANZ ALLDOG A1806

The first version of the Alldog tool carrier tractor from Heinrich Lanz was the A1205, available in 1951 and powered by an 12-hp (8.9-kw) gasoline engine. When the output proved to be inadequate, Lanz introduced a new A1305 version in 1952 equipped with a single-cylinder diesel engine, and this boosted the power output to just 13hp (9.7kw).

Providing one extra horsepower was not exactly extravagant, and it probably made very little difference to the performance in the field, but at least it was an unusual engine. It was air-cooled and was made by Lanz with a single cylinder and a 2,800rpm operating speed, and, although it was designed to run on diesel, the cylinder head was equipped with a spark plug for starting on gasoline.

Final Version
For the final version of the Alldog, known as the A1806 model, Lanz finally got to grips with the power problem, and they chose an MWM diesel engine with liquid cooling and an 18-hp (13.3-kw) output. The first of the A1806 Alldogs were built in 1956, and production continued until the Lanz company was taken over by John Deere. The link with John Deere

Above: *In spite of being under-powered for most of its production life, the Lanz Alldog was easily the most versatile tool carrier available.*

explains why some of the last Alldogs to leave the factory were painted green, although they still bore the Lanz name.

The Alldog is one of the most unusual and versatile designs the tractor industry has produced, and it might have achieved even more success if it had not been seriously underpowered for much of its production life. The tractor was built on a rectangular framework made of steel tubes, with the driver and the engine at one end, and this left most of the framework free to carry a range of implements and attachments.

Equipment

Lanz provided or approved an extraordinary range of equipment to be mounted on, under, in front of, or behind the Alldog. There were more than 50 items covering most of the activities on arable and livestock farms, and the aim was to make the Alldog the universal power unit for virtually every job on the farm. The range included a mid-mounted plow, a manure spreader, a single-row sugar beet harvester, and even a portable milking machine.

All-round visibility from the driver's seat was excellent because of the open frame, but the lack of engine power was not the Alldog's only design fault. Placing the seat so close to the engine must have produced a noise problem for the driver, particularly when working with the air-cooled diesel model.

Above: *With its air-cooled engine within easy reach of the operator's seat, the Lanz Alldog must have been a noisy tractor to drive.*

Specifications

Manufacturer: Heinrich Lanz	**Transmission:** Five-speed gearbox
Location: Mannheim, Germany	**Weight:** N/A
Model: Alldog A1806	**Production started:** 1956
Type: Tool carrier	
Power unit: MWM diesel engine	
Power output: 18hp (13.3kw)	

Below: *This diagram shows the steel rectangle that formed the main frame of the Alldog tool carrier, with space for equipment mounted above and below the frame.*

DOE

⚒ **1958 Ulting, Essex, England**

DOE TRIPLE-D

In the mid-1950s, when a big tractor was 60hp (44.7kw), there were few options available for UK farmers who wanted more horsepower. The obvious choice was a big tracklayer, but in the days of steel tracks there were many farmers who had a strong preference for rubber tires.

This was the problem facing Essex farmer George Pryor. He wanted plenty of power to plow and cultivate his heavy clay soil; the answer, he decided, was to build his own tractor. He bought two Fordson Major tractors and joined them end-to-end, removing the front wheels and axles from both and linking the front of one tractor to the back of the other by an immensely strong turntable that provided the steering action powered by hydraulic rams.

Two Engines

Mr. Pryor's tractor had two engines producing more than 80hp (59kw), two transmissions with four-wheel drive through equal-diameter wheels, and one driver who sat on the seat of the rear tractor operating one set of controls for both power units. Although it looked complicated and awkward, the two-in-one tractor worked well, easily outperforming any of the wheeled tractors available at that time,

Above, top: The hydraulically operated articulation, or hinge point, between the two tractor units provided the steering action for the Triple-D.

Above: This beautifully restored 130hp Doe 130, consisting of two Ford 5000 tractors, is in the Doe company collection.

and the bend-in-the-middle steering provided reasonable maneuverability.

Ernest Doe & Sons, the local Fordson tractor dealer, took an interest in Mr. Pryor's tractor, and it signed an agreement to build an improved version. The first of the Doe tractors was completed in 1958, and it was called the Doe Dual Power, later changed to Doe Dual Drive, then abbreviated to Triple-D.

Implements

The Triple-D attracted considerable interest from farmers and contractors, but a lack of implements suitable for such a powerful tractor meant that Doe & Sons had to design and build plows and cultivators to sell with the tractors. Using Fordson Power Major skid units produced more than 100hp (74kw) from the Triple-D, and switching to Ford 5000 tractor units in 1964 pushed the output to 130hp (96.9kw), and at that stage the tractors' name

was changed to Doe 130. Similarly the Doe 150 produced 150hp (111.8kw) and was based on two Ford 7000s.

By the mid-1960s there was increasing competition from the 100hp (74kw) plus tractors built by the leading tractor companies and by four-wheel drive specialists such as County and Muir Hill. Their big advantage, compared to the Doe tractor, was having just one engine and one transmission to operate and service. This was a significant attraction, and it helped to put the Doe tractor out of production after well over 300 had been built.

Specifications

Manufacturer: Ernest Doe & Sons
Location: Ulting, Essex, England
Model: Doe Triple-D
Type: General purpose
Power unit: Two Ford four-cylinder diesel engines
Power output: 103hp (76.2kw) (Power Major version)

Transmission: Two six-speed gearboxes
Weight: 11,385lb (5,169kg)
Production started: 1958

Below: *A Doe Triple-D tractor is seen here plowing on a field at Ulting, Essex, close to the Doe company headquarters.*

PORSCHE-DIESEL

1957 Friedrichshafen, Germany

PORSCHE-DIESEL JUNIOR L-108

Some of the men who build glamorous and expensive sports cars are also responsible for producing farm tractors. Examples include the Lamborghini car and tractor brands in Italy, David Brown, who owned both the UK tractor company that bore his name, and Aston Martin, and in Germany Dr. Ferdinand Porsche also designed both cars and tractors.

Specifications

Manufacturer: Porsche-Diesel
Location: Friedrichshafen, Germany
Model: Junior L-108
Type: General purpose
Power unit: Single-cylinder diesel engine
Power output: 16hp (11.8kw)
Transmission: Six-speed gearbox
Weight: 2,575lb (1,169kg)
Production started: 1957

Above: *In Germany, as in the United Kingdom, farmers were quick to appreciate the benefits of diesel power, and the Porsche-Diesel Junior featured a single-cylinder diesel engine.*

Dr. Porsche had close links with the Allgaier range of tractors for about 10 years after World War II, and in 1955 the company's name was changed to Porsche. The name change was accompanied by a color change from Allgaier's orange to bright red for Porsche, but the styling remained unaltered.

Smallest Tractor

The smallest tractor in the Allgaier range was the A111 powered by a single-cylinder air-cooled diesel engine, and this model became the Porsche-Diesel Junior L-108 under the Porsche brand name. It had also gained an extra 4hp (2.9kw), apparently achieved by increasing the rated engine speed to 2,250rpm. The single cylinder had a 3.74-in (9.4-cm) bore and 4.56-in (11.5-cm) stroke, and the transmission provided six forward speeds.

The adoption of the Porsche name was followed by an export sales drive with the United States and the United Kingdom as two of the principle targets. Surprisingly it was the Junior model rather than some of the more powerful Porsche tractors that featured strongly in the campaign; however, the lack of power may have deterred some customers, and few Porsche tractors were sold in either country.

INTERNATIONAL HARVESTER

1961 Chicago, Illinois, USA

I.H. GAS TURBINE TRACTOR

Gas turbines were generating a great deal of interest in the late 1950s when most of the leading car manufacturers were experimenting with gas turbine power and it was widely assumed that turbines would replace piston engines for road transportation.

Right: *International Harvester's experimental HT-340 tractor was powered by a gas-turbine engine and featured a hydrostatic transmission with infinitely variable travel speed.*

Specifications

Manufacturer: International Harvester	
Location: Chicago, Illinois	
Model: HT-340	
Type: Experimental	
Power unit: Gas turbine	
Power output: Derated to 40hp (29.8kw)	
Transmission: Hydrostatic	
Weight: N/A	
Produced: 1961	

Air entering the front of a gas turbine is heated by burning fuel, and as it expands it is forced out of the back of the engine at high speed, driving the blades of a turbine in the process. Advantages include compact size and light weight—not particularly useful in a farm tractor—but gas turbines are smoother than a piston engine and more reliable, which would both be welcome. Disadvantages for vehicles include high noise levels, high fuel consumption, and the absence of engine braking.

Hydrostatic Transmission

International Harvester was surprisingly open about these problems in 1961 when it demonstrated its HT-340 experimental tractor. A subsidiary company of International Harvester was developing gas turbines to power helicopters, and one of these was used in the tractor, designed to develop 80hp (59.6kw) but derated to 40hp (29.8kw) for the tractor. The turbine was linked to a hydrostatic transmission, powering a pump to force oil around a circuit to drive motors in the tractor wheels.

The hydrostatic drive was more relevant to future tractor design than the gas turbine. It eliminates the need for a gearbox and provides an infinitely variable range of speeds up to the maximum without altering the engine rpm; hydrostatic drives are also reliable and easy to use. International Harvester later became the leading manufacturer of hydrostatic-drive tractors.

MASSEY-FERGUSON

1964 Coventry, Warwickshire, England, and Detroit, Michigan, USA

MASSEY-FERGUSON MF-135

Project DX was an ambitious program to develop a new range of Massey-Ferguson tractors to be built in the company's American, British, and French factories, and one indication of the size of the project is the fact that the Engineering Division allocated one million man-hours for design work and building and testing the prototypes.

Above: *The Project DX development program produced a new range of Massey-Ferguson tractors, including the popular MF-135 model.*

Left: *MF-135 models were easily Massey-Ferguson's top-selling tractors in the mid-1960s, with production based in Detroit and Coventry.*

The aim of the project was to provide Massey-Ferguson with an all-new range of tractors that were technically advanced and could be marketed on a global scale, and the MF-135 was among the first of the new models. Production started in 1964 at the Detroit factory and in the following year at the Coventry plant. The diesel version was powered by a Perkins engine—Perkins had been taken over by Massey-Ferguson in 1959.

Maximum power output from the three-cylinder diesel engine, recorded in Nebraska in the power-takeoff tests, was 37.8hp (28kw), and the specification included a transmission with 12 forward speeds. American customers were also offered a gasoline-fueled version equipped with a Continental engine.

Top Seller

The MF-135 was the top seller of the MF tractor range in the mid-1960s. Production totaled almost 13,000 tractors at the Detroit factory in 1965, more than one-third of Massey-Ferguson's total U.S. production in that year; the 44,246 MF-135s built in 1966 was well over half of the company's UK output for the year.

Specifications

Manufacturer: Massey-Ferguson
Locations: Coventry, Warwickshire, England, and Detroit, Michigan
Model: MF-135
Type: General purpose
Power unit: Three-cylinder diesel engine
Power output: 37.8hp (28kw)
Transmission: 12-speed gearbox
Weight: 3,645lb (1,655kg)
Production started: 1964

JOHN DEERE
⚒ 1964 Waterloo, Iowa, USA

JOHN DEERE 4020

Another instalment of John Deere's "New Generation of Power" arrived in 1963 when the new top-of-the-range 5010 model was announced, and there were further developments the following year when the 3010 and 4010 models were updated to become the 3020 and 4020.

Right: *The most significant development on the new John Deere 4020 tractor was the powershift transmission, providing easier gear shifting without interrupting the power flow from the engine.*

Specifications

Manufacturer: Deere & Co.

Location: Waterloo, Iowa

Model: 4020

Type: General purpose

Power unit: Six-cylinder diesel

Power out: 91hp (67.3kw)

Transmission: Powershift

Weight: 8,945lb (4,061kg)

Production started: 1964

The biggest difference between the new 3020 and 4020 and their predecessors was the addition of a powershift transmission. This was to prove a significant development because powershifts gradually took over as the preferred transmission for big tractors, but in 1964 they were still an unusual feature.

Advantages of Powershift
Advantages of a powershift include enabling the driver to shift to a different gear ratio smoothly without using the clutch pedal and without interrupting the power flow from the engine to the tractor wheels. This is important in situations where the tractor is working hard, such as pulling a heavy load up a gradient or plowing in difficult conditions, as the tractor loses momentum while the drive is disengaged for a conventional gear change. As well as improving productivity, powershifts also make life easier for the driver, as less physical effort is needed compared with a conventional gear shift.

Other developments featured on the new 4020 model included a hydraulically operated differential lock, using a foot pedal to deliver power equally to both driving wheels in order to reduce the risk of wheelslip, and there was also a power boost to 91hp (63.7kw).

COUNTY

1964 **Fleet, Hampshire, England**

COUNTY SEA HORSE

There is not much demand for a tractor that floats. Nonetheless, County Commercial Cars developed one in 1964, called the Sea Horse. Although it certainly earned plenty of publicity, there is no record of any sales.

The County Sea Horse was based on the 52-hp (38.5-kw) engine and transmission of a Fordson Super Major tractor equipped with a County four-wheel drive conversion. Exactly what is was that prompted the company to produce a floating version of the tractor is not clear, but the development work appears to have been reasonably easy.

Flotation System

Most of the flotation was provided by the large volume of air in the four oversize Goodyear tires, but sealed compartments were also fitted on each wheel to provide additional air volume. County also built large flotation tanks made of sheet steel, and these were mounted on the front and rear of the tractor to provide extra stability in stormy conditions or when passing through the wash of a big ship. The tanks could also be used as ballast containers when the tractor was working on dry land, the company explained.

Apart from the air tanks, the modifications needed to turn the Sea Horse into a marine

Above: *County's Sea Horse tractor photographed during a demonstration as a four-wheel drive tractor working on dry land.*

tractor were surprisingly few. There was no elaborate waterproofing, but the clutch housing was fully sealed, the standard dipsticks were replaced by a screw-in type, and all the transmission housings were equipped with breathers opening above the waterline.

The propulsion system of the Sea Horse was even more straightforward. When the wheels were turning, the tread patterns on the four giant tires acted like paddle wheels, forcing the tractor either forward or backward. The steering mechanism worked in the same way as it would on land, although the turning circle was very much wider.

There were, it seems, just two problems. The propulsion provided by the wheels was much less efficient in the water than on land, and this reduced the forward speed to about 3kts (5.5km/h) in top gear, and the second problem was that the driver of the Sea Horse was soaked by the splashing caused by the wheels.

Channel Crossing
County publicized its latest tractor by taking it for a trip across the English Channel from

France to England, completing the 28 nautical miles (51.8km) in 7 hours and 50 minutes. It also took it to Florida and attracted more publicity when it took the Sea Horse for a trip among the pleasure boats at Fort Lauderdale. If County expected the Sea Horse to sell in large numbers, it was doomed to disappointment, but if, as seems likely, the aim was publicity it was probably delighted.

Left: *An original publicity photograph of the Sea Horse mingling with pleasure boats off the Florida coast.*

Specifications

Manufacturer: County Commercial Cars
Location: Fleet, Hampshire, England
Model: Sea Horse
Type: Amphibious
Power unit: Four-cylinder engine
Power output: 52hp (38.5kw)
Transmission: Six-speed gearbox
Weight: 7,830lb (5,794kg)
Production: 1964

Left: *Another publicity photograph, this time showing the Sea Horse on its way across the English Channel from France to England.*

KUBOTA TALENT 25

The principle reason for Kubota building the Talent 25 tractor was probably to generate some publicity, but the tractor also featured some interesting ideas that the company may have been considering for some of its future production tractors.

Specifications

Manufacturer: Kubota Iron & Machinery Works	
Location: Osaka, Japan	
Model: Talent 25	
Type: Experimental	
Power unit: Diesel engine	
Power output: 25hp (18.5kw)	
Transmission: N/A	
Weight: N/A	
Produced: 1969	

Kubota completed this unique tractor in 1969 and displayed it on its stand at the following year's World Fair in Japan. It featured a number of novel ideas, some of them possibilities for the future, while others were quite obviously impractical.

Closed-Circuit TV

One of the more interesting ideas was using closed-circuit TV cameras to enable the driver to overcome blind spots in his view outside the tractor. The Talent 25 had three cameras linked to a screen inside the cab, and the driver could select which camera he wanted to show images on the screen.

This probably seemed highly futuristic in 1969, but since then some production tractors and self-propelled machines have been fitted with such cameras, usually for safety reasons when the driver is reversing.

Less practical was the shape of the Talent's side panels curving over the tires, which would soon suffer from stone damage and mud splashes. Probably the least practical design feature was the cab with its pair of rear doors for the driver to enter or leave the cab.

Even the most nimble of drivers would find it difficult to climb aboard or make a dignified exit with a large and spiky machine attached to the rear of the tractor. Leading car manufacturers often produce models with futuristic styling only once for their show stands, and these often have more to do with attracting publicity than with assessing reaction to new ideas. Kubota's Talent 25 model is probably also in the same category.

Above: Kubota built the futuristic Talent 25 tractor to exhibit on a show stand in 1969 and to try out new ideas such as the externally mounted CCTV cameras.

LELY

1970 **Maasland, Holland**

LELY HYDRO 90

Lely is best known as one of Europe's leading farm machinery manufacturers, but the company has also experimented with several different tractor projects, although none of them has achieved a lasting commercial impact.

Right: *Lely was one of several companies attracted by the advantages of hydrostatic transmissions.*

Above: *The Hydro 90 model was powered by a six-cylinder MWM engine, and both two and four-wheel drive versions were planned.*

Specifications

Manufacturer: C Van Der Lely NV
Location: Maasland, Holland
Model: Hydro 90
Type: Experimental
Power unit: MWM six-cylinder diesel engine
Power output: 87hp (64kw)
Transmission: Hydrostatic
Weight: 7,062lb (3,210kg) (with four-wheel drive)
Produced: 1970

One of the ideas that attracted the engineers at the Lely company's headquarters in Holland was the idea of using a hydrostatic drive system. This does away with the usual gearbox and clutch, and power is transmitted to the wheels by using a flow of oil driven by a pump and circulating through motors attached to the wheels. The result is a smooth power flow and an infinitely variable travel speed at constant engine rpm.

Engine

Lely's Hydro 90 tractor with hydrostatic drive was demonstrated in 1970. The power unit was a six-cylinder MWM diesel engine with 271.3 cubic inches (4.47 liters) of capacity and an 87-hp (64-kw) output at the rated speed of 2,600rpm. The drive system allowed forward speed to be selected up to a maximum of 12.5mph (20.1km/h), and the top speed in reverse was 7.5mph (12km/h). The tractor was also available in a four-wheel drive version.

As well as the benefits, there is also the major disadvantage with hydrostatic transmissions that they are less efficient than a mechanical gear drive, and this may have been the reason why Lely abandoned the project after building several experimental tractors.

DEUTZ

1972 **Cologne, Germany**

DEUTZ INTRAC 2005

Most of the development work that produced the first "systems" tractors was carried out by German companies during the early 1970s, with Deutz taking the lead with its Intrac tractors.

There were two Intrac models, the 2002 and 2005, and they were both shown for the first time at the DLG agricultural show in 1972. The design of the Intrac followed a new analysis of the way tractors were used on German farms and the equipment that they powered. The analysis also looked at forecasts of the way farming was likely to develop during the next 20 years or so.

Systems Tractors

It was an ambitious project and the result was the Intrac systems tractor, designed to meet the requirements of farmers through the 1970s and beyond. The conclusion reached by the Deutz research team was that tractors would increasingly be used with different combinations of machinery to do a sequence of jobs in a single operation. To achieve this the Intrac was

Above: *When the Intrac models were announced in 1972 they were regarded as the pattern for future systems-tractor development.*

designed to carry both front- and rear-mounted equipment, using an implement linkage and power takeoff at each end of the tractor. The driver's cab was moved to the front of the tractor to leave space for mounting equipment such as a hopper for carrying fertilizer or a tank for spray chemicals.

Other design features of the Intrac included a cab with a large glass area to allow good all-round visibility from the driver's seat, and the transmission for the 2005 version was a hydrostatic drive with two speed ranges, a slower range for field work and a faster range with a maximum speed of 25mph (40km/h) for road travel. Hydrostatic drive was chosen for the Intrac 2005 because it provides infinitely variable adjustment of the forward and reverse travel speed without altering the engine speed and power output.

Four-Wheel Drive

Four-wheel drive was standard on the Intrac 2005, using equal-sized front and rear wheels,

but only an option on the smaller 2002 version, which also had smaller-diameter front wheels and a mechanical transmission. Power for both models was provided by a Deutz air-cooled engine providing 90hp (66.6kw) for the 2005 and 51hp (38kw) for the 2002.

The Intracs attracted enormous interest, and they were widely regarded as the first stage in a new revolution in tractor design. They and some of the other early examples of systems tractors were certainly influential, and some of the features of the early systems models are still available on some current production tractors. The Intracs did not achieve the sales volumes that were widely predicted, however, and in commercial terms they were not a big success.

Specifications

Manufacturer: Klöckner/Humboldt-Deutz AG	**Power output:** 90hp (66.6kw)
Location: Cologne, Germany	**Transmission:** Two-range hydrostatic
Model: Intrac 2005	**Weight:** N/A
Type: Systems tractor	**Production started:** 1972
Power unit: Five-cylinder diesel engine	

Below: *A mid-mounted cab plus attachment points for front- and rear-mounted equipment gave the Intrac exceptional versatility.*

VALMET

🛠 **1975 Jyväskylä, Finland**

VALMET 1502

There is plenty of evidence that four-wheel drive provides more efficient traction than two-wheel drive when the ground surface is wet and slippery. Similar information about the effectiveness of six-wheel drive, however, is not as widely available.

Specifications

Manufacturer: Valmet Oy
Location: Jyväskylä, Finland
Model: 1502
Type: General purpose
Power unit: Six-cylinder diesel engine
Power output: 136hp (100kw)
Transmission: 16-speed gearbox
Weight: 16,434lb (7,470kg)
Production started: 1975

Above: *In 1975 the Valmet company in Finland, now known as Valtra, introduced the 136-hp (100-kw) 1502 model with six-wheel drive.*

The Valmet company in Finland presumably decided that the extra cost of six-wheel drive could be justified, and it announced its first—and also its last—six-wheel drive tractor in 1975. The front end of the Valmet 1502 tractor was conventional, with a single axle and two driving wheels; however, the rear end was supported on a bogie unit with a set of four close-coupled powered wheels.

Six-Wheel Drive

Valmet engineers designed the rear bogie unit to allow plenty of vertical movement of both the front and the rear sections, and the company claimed that this ensured efficient traction over rough ground, with a smoother ride for the operator. An extra benefit was reduced soil compaction as the weight of the tractor was spread over six wheels instead of the usual four. One of the options demonstrated by the company was a steel track unit over the two rear wheels on each side, allowing a further improvement in both the traction and the low ground-pressure characteristics of the tractor.

Valmet designed the 1502 tractor to suit the industrial and forestry markets as well as the farming industry, but sales of the model were disappointing, and customers were clearly not convinced that the extra cost of six-wheel drive was justified by the benefits.

LELY

✘ 1977 Temple, Texas, USA

LELY MULTIPOWER

One of the surprise competitors in the high-horsepower sector of the North American tractor market was the Lely company from Holland, and it also produced one of the most unconventional designs.

Above: When European machinery specialist Lely decided to break into the American high-horsepower market it built the 420-hp (311-kw) Multipower with two engines.

Specifications

Manufacturer: C Van Der Lely NV

Location: Temple, Texas

Model: Multipower (later called 420)

Type: General purpose

Power unit: Two Caterpillar V-8 diesel engines

Power output: 420hp (311kw)

Transmission: Two 10-speed gearboxes

Weight: 35,000lb (15,890kg) (shipping weight)

Production started: circa 1977

As one of Europe's leading manufacturers of power takeoff driven machines, Lely wanted to expand the sales of its equipment in the United States and Canada, but one of the problems in the 1970s was the absence of a three-point linkage and power takeoff drive on the biggest tractors.

Twin Engines

The Lely company decided to build its own high-horsepower tractor to work with power takeoff equipment, and it assembled a team of engineers at its American distribution center in Texas to design and build one or more prototype tractors. Presumably it was an expensive project, but it did not run smoothly. The design included articulated steering and two engines producing 210hp (155.5kw) each.

The engine in the front section drove the front wheels and could be used on its own to reduce fuel consumption when full power was not needed, while the rear engine powered the back wheels and also provided power for the power takeoff shaft.

The twin engine idea, which also involved two transmissions and two fuel tanks, was expensive and put the Multipower at a cost and weight disadvantage. When the tractor was launched, Massey-Ferguson threatened legal action because it had already registered the Multipower name, and Lely had to back down and call the tractor the Lely 420 instead.

It is not known if any Multipower or 420 tractors were sold, but we do know that Lely soon abandoned the project to concentrate on its machinery business.

VERSATILE

1977 Winnipeg, Manitoba, Canada

VERSATILE 1080 "BIG ROY"

By the mid-1970s competition in the high-horsepower sector of the tractor market was intense, and there was rivalry to offer the most powerful tractor. The Canadian-based Versatile company was one of the leading manufacturers of big four-wheel drive tractors and Roy Robinson, its high-profile managing director, who stood at 6 ft 4 in (1.93 m) tall, decided to put his company ahead in the power race by building the biggest tractor in the world.

In the mid-1970s the biggest tractors were about 350hp (260kw), but Versatile engineers decided to use a Cummins engine producing 600hp (444kw) from 1153.3-cubic-inches (19-liters) capacity. In terms of tractor design this was venturing into unknown territory, and there were concerns about tire damage when transmitting so much engine power, and also the soil damage that might be caused by so much weight on just four wheels and tires.

Articulated Tractor

To overcome these problems the design team used four axles and an eight-wheel drive layout, and the tractor was in two halves with a pivot point in the middle to provide hydraulically operated articulated steering. The tractor was completed in 1977 and it was called "Big Roy" in honor of Mr. Robinson.

The weight of the finished tractor was 26.4 tons (26.8 tonnes), and the overall length was

Above, top: *The Canadian-based Versatile company named its most powerful tractor Big Roy after Roy Robinson, its managing director.*

Above: *A CCTV camera on the back of Big Roy was used when reversing, as the 600-hp (444-kw) engine was mounted behind the cab where it blocked the rear view.*

33ft (10m). It was—and still is—probably the most impressive-looking tractor ever built.

A ladder on each side provided access to the cab, and inside the cab there was plenty of space for the driver and a passenger. The cab was air-conditioned, a rare luxury when Big Roy was built, and there was also a small screen connected to a closed-circuit TV camera at the rear of the tractor. The rear view from the driver's seat was totally blocked by the engine compartment mounted high up on the rear section of Big Roy, and the TV equipment was essential when reversing and to position the tractor correctly when hitching up to an implement.

When the tractor was completed it was taken to some of the leading American and Canadian agricultural shows, where it attracted considerable publicity, and there were plenty of big-acreage farmers who were interested in buying a 600-hp (444-kw) tractor.

Problems

Big Roy was taken on a demonstration tour, and this is where the problems emerged. There were virtually no implements big enough for a tractor of this size, and excessive damage to the tire lugs and casing was caused by the steering action. Development work on Big Roy was abandoned, and was replaced by a more modest project to build a 470-hp (350.4-kw) tractor that became the Versatile 1150. Big Roy found a new home in the Agricultural Museum in Brandon, Manitoba, where it is one of the main attractions.

Specifications

Manufacturer: Versatile Manufacturing

Location: Winnipeg, Manitoba, Canada

Model: 1080 Big Roy

Type: Pulling tractor

Power unit: Cummins KTA six-cylinder diesel engine

Power output: 600hp (444kw)

Transmission: Six-speed gearbox

Weight: 57,580lb (26,141kg)

Produced: 1977

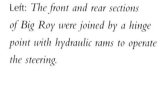

Left: *The front and rear sections of Big Roy were joined by a hinge point with hydraulic rams to operate the steering.*

MASSEY-FERGUSON

1978 Des Moines, Iowa, USA

MASSEY-FERGUSON 4880

The 4000 series tractors took Massey-Ferguson into the high-horsepower sector of the market, with three models providing outputs ranging from 180 to 273hp (134.2 to 202kw) measured at the power takeoff.

All three models were powered by Cummins V-8 engines, but the engine in the 4880 model at the top of the range included a turbocharger to boost the output. The cylinder dimensions were 5.5-in (13.9-cm) bore and 4.75-in (12-cm) stroke, and the semipowershift transmission provided 18 forward gears and a top speed of 19.2mph (30.8km/h).

With plenty of engine power, four-wheel drive with dual wheels all round plus articulated steering, the 4000 series had all the characteristics of the typical giant North American pulling tractor, but they were equipped with a three-point linkage and a power takeoff to increase their versatility. They were also the first production tractors with an electronic control system for the rear linkage. Electronics were becoming familiar on tractors and combine harvesters by the late 1970s, but mainly for collecting and displaying information from sensors that monitored factors such as the speed of drive shafts or the engine cooling system's temperature.

Electronic Breakthrough

The electronics on the 4000 series tractors represented a breakthrough in technology because they collected data from sensors measuring the pulling force on the linkage arms and the implement depth, and they used the information to adjust the linkage position automatically, providing faster, more accurate control than even an experienced tractor driver.

Above: *Massey-Ferguson's 4000 series high-horsepower models were the first tractors with an electronic control system for the rear linkage.*

Specifications

Manufacturer: Massey-Ferguson
Location: Des Moines, Iowa
Model: 4880
Type: General purpose
Power unit: V-8 turbo diesel engine
Power output: 273hp (202kw)
Transmission: 18-speed semipowershift
Weight: 31,090lb (14,115kg)
Production started: 1978

DAVID BROWN/CASE

⚒ **1980   Meltham, Yorkshire, England**

DAVID BROWN/ CASE 1290

David Brown Tractors produced one of Britain's most successful tractor ranges, but financial problems in the early 1970s lead to a successful takeover bid by Tenneco, the U.S.-based company that already owned Case and later acquired International Harvester.

Right: *Like other British-built 90 series tractors, the 1290 model was sold under the David Brown name in the United Kingdom and some export markets.*

Above: *All American- and British-built 90 series tractors sold in the United States carried the Case brand name.*

Specifications

Manufacturer: David Brown Tractors

Location: Meltham, Yorkshire, England

Model: 1290

Type: General purpose

Power unit: Four-cylinder diesel engine

Power output: 54hp (40kw) (at the power takeoff)

Transmission: 12-speed gearbox

Weight: 6,570lb (2,983kg)

Production started: 1980

The new owners decided to continue to feature the David Brown name for several years. David Brown had many loyal customers and dealers, and when the new 90 series tractors were launched in 1980 they still carried the David Brown name and badge in most markets including the United Kingdom; the Case name was used for those exported to the United States.

End of the Line

The 90 series was the result of a coordinated development program, with the Case factory at Racine building the high-horsepower models that were introduced in 1978, while the small to medium tractors were built at the David Brown plant. The 1290 was one of the British-built tractors, powered by a four-cylinder engine that delivered 54hp (40kw) at the power takeoff when it was tested under the Case brand name at Nebraska.

Production of the British-built 90 series tractors came to an end when they were replaced by the 94 series, and the bad news for David Brown enthusiasts was that this was also the end of the line for David Brown tractors, as the new 94 models all carried the Case badge.

CATERPILLAR
✖ 1987 Peoria, Illinois, USA

CATERPILLAR CHALLENGER 65

Tracklaying tractors are unbeatable for turning engine power into drawbar pull when the ground conditions are wet and slippery, and they are also unbeatable for spreading the weight of the tractor over a larger surface area to reduce soil compaction. The traditional steel tracks do, however, have some serious disadvantages.

Steel tracks are noisy, they are not suitable for fast travel speeds, and in most countries they are not allowed to drive on public roads because of the damage they cause. Sales of tracklayers remained high while the only alternative was a two-wheel drive tractor, but as more manufacturers offered four-wheel drive models from the 1950s onward sales of crawler tractors fell sharply.

By the early 1980s it seemed increasingly likely that crawler tractors would soon be restricted to niche markets such as farms with very steep land, where the tracks give extra stability, but then, in 1987, the situation changed as the Caterpillar Challenger 65 tractor arrived with its steel-reinforced rubber tracks.

New Tracks
The new tracks—Caterpillar call them the Mobil-trac system—were probably the most important technical development in crawler

Above, top: *Rubber tracks arrived on the Caterpillar Challenger 65, bringing a halt to the steady decline in crawler-tractor sales.*

Above: *Claas took over the marketing of Caterpillar Challenger tractors in the United Kingdom and other European countries.*

tractor design for 50 years or more. The rubber-tracked Challenger 65 and the models that followed retain the traditional tracklayer benefits of efficient pulling power and low ground pressure, but they can also be used on public roads, they can travel at the same speed as an ordinary wheeled tractor, and the driver benefits from much lower noise levels than conventional steel tracks.

Reports from Challenger users showed that the tracks could compete with rubber tires in terms of strength, durability, and replacement cost on tractors of similar horsepower, and Challenger sales increased. The success of the Challenger 65 and subsequent Caterpillar models encouraged more tractor manufacturers to offer their own rubber-tracked models, reversing the previous downward trend in crawler tractor sales.

Engine

The Challenger 65, the tractor that introduced the benefits of rubber tracks, was powered by a Caterpillar engine with turbocharger and intercooler. The engine power was 270hp (200kw), delivered through a powershift transmission with 10 forward speeds. Driver-comfort features included full air conditioning and an air-cleaning system with a dust extractor. There was also a cigarette lighter and an ashtray. Steering was controlled by a steering wheel with a differential control system that speeded up one track and slowed down the other.

Having achieved so much success with its Challenger tractors, Caterpillar caused considerable surprise in the industry when it decided to pull out of the agricultural equipment market in 2002, selling the Challenger tractor business to Agco.

Specifications

Manufacturer: Caterpillar
Location: Peoria, Illinois
Model: Challenger 65
Type: Pulling tractor
Power unit: Six-cylinder engine
Power output: 270hp (200kw)
Transmission: 10-speed powershift
Weight: 31,000lb (14,060kg)
Production started: 1987

Below: *The Challenger's rubber tracks offered the pulling efficiency of traditional steel tracks, but without the restrictions on road travel and forward speed.*

JCB

✗ **1991   Cheadle, Staffordshire, England**

JCB FASTRAC 145

During the 1970s and 1980s a large number of familiar names vanished from the tractor industry in the United States and Europe as a result of mergers, takeovers, and bankruptcies, and this is a continuing trend. It is more unusual for a big company to move into tractor production for the first time, but this happened in 1991 when JCB introduced the Fastrac tractor range.

JCB is a world leader in the production of backhoe loaders and other construction machinery, and it also claims to be the biggest manufacturer of telescopic loaders for the farming industry. Even for a big company such as JCB, trying to challenge the established names in the tractor industry with just another conventional tractor would have been extremely difficult, but instead they chose a highly unconventional design.

High Speed

The Fastrac was described in the original JCB press release as "the world's first genuine high-speed tractor." The first production models were each capable of pulling a 13.7-ton (14-tonne) load at more than 40mph (64km/h) on the road, almost twice the top speed of most conventional tractors, but the Fastrac's four-wheel drive through equal-diameter front and rear wheels was also designed to provide

Above, top: The move by JCB, a leading manufacturer of construction equipment, into the overcrowded tractor market was a surprise.

Above: Although the Fastrac's road performance dominated much of the original publicity, the tractors were also designed for heavy draft work such as plowing.

efficient slow-speed traction for jobs such as plowing.

The first two Fastrac models were the 125 and the 145, with the model number in each case indicating the horsepower. All Fastracs are equipped with Perkins engines, with a turbocharger added to the engine of the 145 model to boost the power output, and the transmission for both models provided 18 forward speeds and six in reverse.

Fastrac Suspension

Tractors designed for such a high transportation speed must have plenty of braking power, and the Fastracs were equipped with a truck-type braking system with air-operated disks on all four wheels. Another design feature to help cope with the fast travel speeds was a full suspension system on both the front and the rear wheels, helping to smooth out the bumps and provide increased stability to improve the steering control.

A special feature designed into the Fastrac suspension—still used on the latest models—is the self-leveling action at the rear. This is essential for jobs such as plowing where it is important to maintain a constant depth while carrying the weight of a mounted plow on the rear linkage.

Another Fastrac feature is a load space behind the cab with of capacity of 2.4 tons (2.5 tonnes), often used to carry crop-spraying equipment, and the options list on all models includes a factory-fitted front linkage.

The first Fastracs were built at the main JCB plant, but production was transferred to a separate factory operated by JCB Landpower, the agricultural equipment arm of JCB.

Specifications

Manufacturer:	JCB Landpower
Location:	Cheadle, Staffordshire, England
Model:	Fastrac 145
Type:	General purpose
Power unit:	Six-cylinder turbo diesel engine
Power output:	145hp (107kw)
Transmission:	18-speed gearbox
Weight:	13,831lb (6,274kg)
Production started:	1991

Below: *Design features of the Fastrac include a front and rear suspension system, four-wheel braking, and a load platform behind the cab.*

CLAAS

1997 **Harsewinkel, Westphalia, Germany**

CLAAS XERION 2500

When its joint project with Schluter to develop the Eurotrac tractor did not result in a production tractor, Claas adopted the do-it-yourself approach and developed its own systems tractor. It was called the Xerion, and it was the most ambitious systems tractor produced so far, requiring substantial financial and technical resources.

Above: *With equipment on the front and rear linkages and on the load platform, the Xerion was the newest addition to the list of systems tractors.*

Claas is Europe's biggest and most successful manufacturer of harvesting machinery, but it has also had ambitions to move into the tractor market, and the Eurotrac and the Xerion were steps toward achieving this aim.

Systems Tractor

The Xerion followed the usual systems-tractor layout of four-wheel drive through equal-diameter wheels plus attachment points for equipment at the front and the rear of the tractor. Three power-takeoff points were provided at the front, rear, and middle of the tractor, and there was also space to carry equipment on the load area over the rear section. The Xerion's versatility as a power unit went much further than previous systems tractors, however, and it was designed to operate complex wraparound equipment such as root-harvesting machines which, in some cases, almost obscured the tractor unit.

Instead of a reversible driving seat and controls, the complete cab on the Xerion could swivel hydraulically to face forward or to the rear, and there were also two positions for the cab, one at the front and the other near the middle of the tractor, with hydraulic power to switch positions.

Transmission

Less eye-catching, but with much more long-term significance, was the transmission developed by Claas for the new tractor. It was based on a mechanical gearbox with eight ratios, plus a separate hydrostatic transmission. The hydrostatic drive allowed stepless travel-speed adjustment in each of the eight gears without varying the speed of the engine, and transmissions of this type would soon be appearing elsewhere in the tractor industry.

The steering system on the Class Xerion was also unconventional. It was capable of being operated through all four wheels, and the driver was able to select from three different steering modes.

Claas demonstrated a prototype version of the Xerion in 1993; after further development work the first production tractors arrived in 1997. They were available in versions of 200, 250, and 300hp (149.1, 185, and 233.7kw), powered by six-cylinder engines equipped with a turbocharger and intercooler. They were aimed at big-acreage farms and the largest agricultural contractors, who form the exclusive top end of the tractor market in Europe.

Specifications

Manufacturer: Claas
Location: Harsewinkel, Westphalia, Germany
Model: Xerion 2500
Type: Systems tractor
Power unit: Six-cylinder engine
Power output: 250hp (185kw)
Transmission: Constantly variable
Weight: N/A
Production started: 1997

Above: *This Xerion tractor is almost hidden under a specially designed wraparound sugar-beet harvester.*

Left: *The Xerion was the first tractor designed with a cab that could be moved hydraulically to the middle or front to suit the job it is undertaking.*

JOHN DEERE
⚒ 1997 Mannheim, Germany

JOHN DEERE 6910

When John Deere introduced its new 6010 series tractors in 1997, one of the sales features was the new Triple Link Suspension, or TLS, available as an option on models from 100hp (74.5kw) upward.

Above: *John Deere 6010 series tractors were among the first to be offered with a suspension system on the front axle.*

Specifications

Manufacturer: Deere & Co.	
Location: Mannheim, Germany	
Model: 6910	
Type: General purpose	
Power unit: Six-cylinder engine	
Power output: 135hp (100kw)	
Transmission: 24-speed semipowershift	
Weight: 10,450lb (4,750kg)	
Production started: 1997	

B y the late 1990s there were still just a tiny minority of tractor manufacturers offering any form of suspension system, but they were attracting increased interest as farmers and contractors understood the importance of providing a smoother, more comfortable ride for the driver. By 1997 most of the leading tractor manufacturers were developing suspension systems for the front axle or the cab; however, it was the John Deere company that sparked renewed interest in front-axle suspension.

Suspension System

TLS uses a combination of hydraulic cylinders and gas-filled accumulators to absorb some of the bounce and vibration from the front

wheels. As well as giving the driver a smoother ride, this produces more stability and better steering control when traveling at speed, and it is also said to improve contact between the front wheels and the ground to produce better traction when working with four-wheel drive.

The first 6010 series tractors covered the big-selling 75- to 135-hp (55.9- to 100-kw) sector of the tractor market and were the most popular models in the John Deere range. The John Deere 6910 model was the most powerful model in the series, with 135hp (100kw) available from a six-cylinder engine, although more powerful models were introduced later when the 6010 models were upgraded to become the 6020 series.

Above, left: *The German-built 6010 series tractors from John Deere covered the medium-power sector of the market and were the company's top-selling models.*

BUHLER
🔧 2000 Winnipeg, Manitoba, Canada

BUHLER VERSATILE 2425

Below: The Buhler Versatile 2425 is built at the same plant in Winnipeg, Manitoba, that had previously built Versatile and New Holland tractors.

When the New Holland group was forced to sell some of its factories in the late 1990s, one of the disposals was the tractor plant in Canada, and the buyer was the Buhler company.

Specifications

Manufacturer: Buhler Industries
Location: Winnipeg, Manitoba, Canada
Model: Buhler Versatile 2425
Type: Pulling tractor
Power unit: Six-cylinder Cummins engine
Power output: 425hp (314kw)
Transmission: 12-speed gearbox
Weight: N/A
Production started: 2000

The Winnipeg factory was originally owned by Versatile, a leading manufacturer of high-horsepower tractors, and Ford bought Versatile to secure its supply of big tractors. New Holland was forced to sell the factory, plus several of its European factories, in order to gain approval under antitrust legislation for merging its tractor and farm machinery business with those of Case I.H. to form CNH.

Versatile Range
Under New Holland ownership the product range built in the Winnipeg factory had included the high-horsepower models with articulated steering, the bi-directional tractor with a reversible driving position, and New Holland's Genesis tractor range. Under the terms of the sale agreement the new owners were able to continue building these models.

The biggest of the red-painted Buhler Versatile tractors is the high-horsepower 2425, featuring four-wheel drive and usually equipped with dual wheels and tires. The power unit is a six-cylinder Cummins engine with a turbocharger and intercooler, and equipped with full electronic management, and the rated output is 425hp (314kw) from 849.8 cubic inches (14 liters) of capacity. A 12-speed QuadShift transmission provides four synchro gears in each of three forward ranges.

CASE IH
✗ 2003   Basildon, Essex, England

CASE I.H. JX1100U MAXXIMA

Case I.H. has introduced a new range of tractors in the 70- to 100-hp (52.1- to 74.5-kw) market sector and equipped them with a new four-cylinder engine developed jointly by three leading engine manufacturers.

Specifications

Manufacturer: Case I.H.
Location: Basildon, Essex, England
Model: JX1100U Maxxima
Type: General purpose
Power unit: Four-cylinder engine
Power output: 100hp (74.5kw)
Transmission: 24-speed powershift with power shuttle
Weight: N/A
Production started: 2003

Behind the new engine is a research and development program financed jointly by Case I.H., Cummins, and Fiat's Iveco engine subsidiary. They operated from a special research center in the United Kingdom, and the 273.1-cubic-inch (4.5-liter) power unit for the new Case tractors was one of the results of what was known as the European Engine Alliance.

Engines
All four of the new JXI-U Maxxima tractors are equipped with basically the same engine, which is naturally aspirated for the 72- and 82-hp (53.6- and 61.1-kw) models and equipped with different levels of turbocharging for the

91- and 100-hp (67.8- and 74.5-kw) versions. Advantages claimed for the new engines include reduced wear rates to minimize oil consumption, fuel consumption has improved by 5 percent compared to the previous engines, and measures to reduce engine-noise levels have achieved a 1.5db(A) reduction.

All models are available in two- or four-wheel drive versions, and there is a choice of three different transmission options. The top-priced option is a 24-speed powershift with a power shuttle. The power shuttle, operated by a control mounted on the steering column, allows the driver to change between forward and reverse without using the clutch pedal.

Above: *The four-cylinder engine powering the four-model JXU Maxxima models from Case I.H. is a new design developed jointly by Case, Cummins, and the Fiat group's Iveco engine operation.*

NEW HOLLAND
✗ 2004   Basildon, Essex, England

NEW HOLLAND TVT 190

It was no surprise when New Holland joined the rapidly growing list of manufacturers offering a tractor with a continuously variable transmission or CVT. It happened in 2004 when New Holland launched the new TVT series.

Right: *One of the recent converts to constantly variable transmission (CVT) drive systems is New Holland, with its TVT series tractors announced in 2004.*

Specifications

Manufacturer: New Holland

Location: Basildon, Essex, England

Model: TVT190

Type: General purpose

Power unit: Iveco six-cylinder turbo engine

Power output: 192hp (142kw)

Transmission: Constantly variable

Weight: N/A

Production started: 2004

There are five TVT models with power outputs ranging from 137 to 192hp (102.1 to 142kw), and they are powered by high-specification Iveco engines with six cylinders and 400.6 cubic inches (6.6 liters) of capacity. The engines are turbocharged and intercooled, and all feature electronic injection management.

Engine Options

The options list for the engines includes a cooling fan with a variable blade angle that is automatically adjusted by a thermostatic control system. An increase in the temperature of the engine adjusts the blade angle to move a greater volume of air and increase the cooling action. The blade angle is kept to a minimum while the engine is running at less than the optimum temperature after a cold start, as this reduces the time taken for the warming-up process when the engine is running at less than its peak efficiency.

Like other CVTs, New Holland's transmission on the TVT tractors consists of both mechanical and hydrostatic drive systems operating in three speed ranges. The ranges cover creeper speeds, a middle range for field work, and a faster transportation range for road travel. There is also a power shuttle for push-button changing between forward and reverse without using the clutch pedal, and the computerized control system can be set to adjust the engine speed and transmission automatically in order to maintain a constant forward speed when working with equipment operated using power-takeoff.

Bibliography

Ashby, J. E. *British Tractors & Power Cultivators.*
Eastbourne, UK: Pentagon Publications, 1949.

Bell, Brian. *Fifty Years of Farm Tractors.* Ipswich, UK:
Farming Press, 1999.

Directory of Wheel & Track-Type Tractors.
Rome: UN Food & Agricultural Organization, 1955.

Fraser, Colin. *Harry Ferguson Inventor and Pioneer.*
London: John Murray, 1972.

Gibbard, Stuart. *Ford Tractor Conversions.* Ipswich, UK:
Farming Press, 1995.

Gibbard, Stuart. *The Ferguson Tractor Story.*
Ipswich, UK: Old Pond Publishing, 2000.

Gibbard, Stuart. *The Ford Tractor Story 1917–1964.*
Ipswich, UK: Old Pond Publishing & Japonica Press, 1998.

Gibbard, Stuart. *The Ford Tractor Story 1964–1999.*
Ipswich, UK: Old Pond Publishing & Japonica Press, 1999.

Hafner, Kurt. *Lanz von 1928 bis 1942.*
Stuttgart: Franckh Historische Tecnik, 1989.

Hafner, Kurt. *Lanz von 1942 bis 1955.* Stuttgart:
Franckh-Kosmos, 1990.

Macmillan, Don. *John Deere Tractors and Equipment Volume I.*
St. Joseph, Michigan: American Society of Agricultural
Engineers, 1988.

Two Cylinder Collector Series Volume II. Grundy Center, Iowa:
Two Cylinder Club, 1993.

Wendell, C. H. *American Farm Tractors.* Saratosa, Florida:
Crestline Publishing, 1979.

Wendell, C. H. *International Harvester.* Saratosa, Florida:
Crestline Publishing, 1993.

Wendell, C. H. *Nebraska Tractor Tests Since 1920.* Saratosa,
Florida: Crestline Publishing, 1985.

Wik, Reynold M. *Henry Ford & Grassroots America.*
Ann Arbor, Michigan: University of Michigan Press, 1972.

Williams, Michael. *Great Tractors.* Poole, Dorset: Blandford
Press, 1982.

Williams, Michael. *Classic Farm Tractors.* Poole, Dorset:
Blandford Press, 1984.

Williams, Michael. *Ford & Fordson Tractors.* Poole, Dorset:
Blandford Press, 1985.

Williams, Michael. *Massey-Ferguson Tractors.* Poole, Dorset:
Blandford Press, 1987.

Williams, Michael. *John Deere Two-Cylinder Tractors.*
Suffolk, UK: Farming Press, 1993.

Williams, Michael. *Farm Tractors.* Guilford, Connecticut:
The Lyons Press, 2002

Index

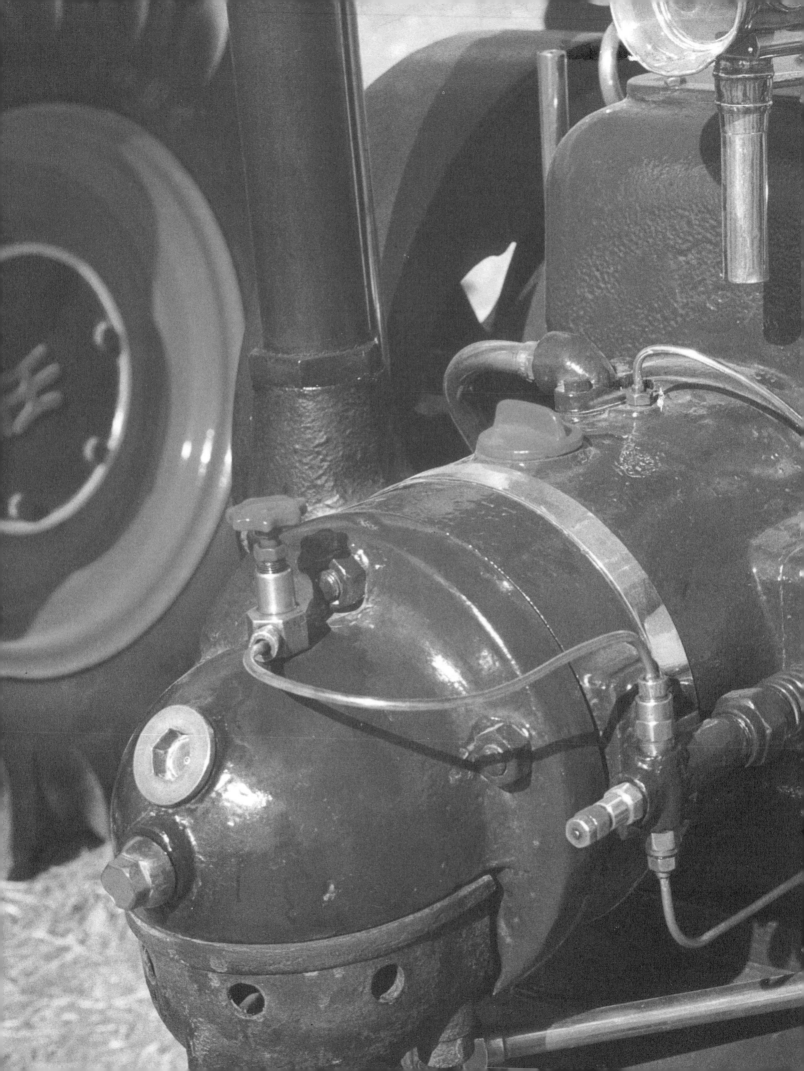